NATIONAL GEOGRAPHIC

World Cultures and Geography

SOUTHEAST ASIA

GEO

Go interactive with myNGconnect.com

Acknowledgments

Grateful acknowledgment is given to the authors, artists, photographers, museums, publishers, and agents for permission to reprint copyrighted material. Every effort has been made to secure the appropriate permission. If any omissions have been made or if corrections are required, please contact the Publisher.

Photographic Credits

Front Cover: (bkg) © Scott Kemper; (front) Alamy, © Lonely Planet Images/Getty Images

Acknowledgments and credits continued on page RB116.

Visit National Geographic Learning online at www.NGSP.com

Visit our corporate website at www.cengage.com

Printed in the USA
RR Donnelley
Menasha, WI

ISBN: 978-07362-9011-1

13 14 15 16 17 18 19 20

10 9 8 7 6 5 4 3

NATIONAL GEOGRAPHIC

World Cultures and Geography OVERVIEW

STUDENT EDITION UNITS	COMPREHENSIVE (SURVEY)	EASTERN HEMISPHERE	WESTERN HEMISPHERE
The Essentials of Geography	•	•	•
North America	•		•
Central America & the Caribbean	•		•
South America	•		•
Europe	•	•	•
Russia & the Eurasian Republics	•	•	•
Sub-Saharan Africa	•	•	
Southwest Asia & North Africa	•	•	
South Asia	•	•	
East Asia	•	•	
Southeast Asia	•	•	
Australia, the Pacific Realm & Antarctica	•	•	

UNIT 11 Southeast Asia

TEACHER'S EDITION

CONSULTANTS AND REVIEWERS

Program Consultants

Peggy Altoff
District Coordinator
Past President, NCSS

Mark H. Bockenhauer
Professor of Geography,
St. Norbert College

Andrew J. Milson
Professor of Social Science Education
and Geography, University of Texas (Arlington)

David W. Moore
Professor of Education,
Arizona State University (Phoenix)

Janet Smith
Associate Professor of Geography,
Shippensburg University

Michael W. Smith
Professor, Department of Curriculum,
Instruction, and Technology in Education,
Temple University

Teacher Reviewers

Kayce Forbes
Deerpark Middle School
Austin, Texas

Michael Koren
Maple Dale School
Fox Point, Wisconsin

Patricia Lewis
Humble Middle School
Humble, Texas

Julie Mitchell
Lake Forest Middle School
Cleveland, Tennessee

Linda O'Connor
Northeast Independent School District
San Antonio, Texas

Leah Perry
Exploris Middle School
Raleigh, North Carolina

Robert Poirier
North Andover Middle School
North Andover, Massachusetts

Heather Rountree
Bedford Heights Elementary
Bedford, Texas

Erin Stevens
Quabbin Regional Middle/High School
Barre, Massachusetts

Beth Tipper
Crofton Middle School
Crofton, Maryland

Mary Trichel
Atascocita Middle School
Humble, Texas

Andrea Wallenbeck
Exploris Middle School
Raleigh, North Carolina

Reviewers of Religious Content

Charles Haynes
First Amendment Center
Washington, D.C.

Shabbir Mansuri
Institute on Religion and
Civic Values
Fountain Valley, California

Susan Mogull
Institute for Curriculum Reform
San Francisco, California

Raka Ray
Chair, Center for South Asia Studies
University of California
(Berkeley)

NATIONAL GEOGRAPHIC EXPLORERS, FELLOWS, AND GRANTEES

Greg Anderson
National Geographic Fellow

Katey Walter Anthony
National Geographic Emerging Explorer

Ken Banks
National Geographic Emerging Explorer

Katy Croff Bell
National Geographic Emerging Explorer

Christina Conlee
National Geographic Grantee

Alexandra Cousteau
National Geographic Emerging Explorer

Thomas Taha Rassam (TH) Culhane
National Geographic Emerging Explorer

Jenny Daltry
National Geographic Emerging Explorer

Wade Davis
National Geographic Explorer-in-Residence

Sylvia Earle
National Geographic Explorer-in-Residence

Grace Gobbo
National Geographic Emerging Explorer

Beverly Goodman
National Geographic Emerging Explorer

David Harrison
National Geographic Fellow

Kristofer Helgen
National Geographic Emerging Explorer

Fredrik Hiebert
National Geographic Fellow

Zeb Hogan
National Geographic Fellow

Shafqat Hussain
National Geographic Emerging Explorer

Beverly Joubert
National Geographic Explorer-in-Residence

Dereck Joubert
National Geographic Explorer-in-Residence

Albert Lin
National Geographic Emerging Explorer

Elizabeth Kapu'uwailani Lindsey
National Geographic Fellow

Sam Meacham
National Geographic Grantee

Kakenya Ntaiya
National Geographic Emerging Explorer

Johan Reinhard
National Geographic Explorer-in-Residence

Enric Sala
National Geographic Explorer-in-Residence

Kira Salak
National Geographic Emerging Explorer

Katsufumi Sato
National Geographic Emerging Explorer

Paola Segura
National Geographic Emerging Explorer

Beth Shapiro
National Geographic Emerging Explorer

Cid Simoes
National Geographic Emerging Explorer

José Urteaga
National Geographic Emerging Explorer

Spencer Wells
National Geographic Explorer-in-Residence

Best Practices For ACTIVE Teaching

To bring best practices into your classroom, choose from the following technology components and instructional routines and make them part of your daily instruction. Many of these practices are also built directly into the instruction in your Teacher's Edition.

PROGRAM TECHNOLOGY

World Cultures and Geography provides a variety of technology to make your job easier and to help students become motivated, independent learners. Use the online components listed below to supplement print resources or to create an entirely digital learning environment.

STUDENT COMPONENTS	TEACHER COMPONENTS
Student eEdition	Teacher's eEdition
Interactive Map Tool	Core Content Presentations
Digital Library	Online Lesson Planner
Magazine Maker	Assessment
Connect to NG	Interactive Whiteboard GeoActivities
Maps and Graphs	Guided Writing
GeoJournal	Teacher Resources
Student Resources	

All digital resources and more information about them are available at **myNGconnect.com**. In addition, see the next page for specific instructions on how to use the **Interactive Map Tool**—a technology component that plays an integral part in the Teacher's Edition instruction.

Interactive Map Tool

The **Interactive Map Tool** is an online mapmaker that allows students to draw and add labels and data layers to a map. For general use, follow the instructions below to navigate the map tool.

1. Access the tool at **myNGconnect.com**.
2. Select a region to explore. Click on the "Region" menu to select a continent and the "Country" menu to select a specific country, or simply click and drag the map to maneuver it to a specific location. You can also use the slider bar to zoom in and out.
3. Choose the type of map you wish to view by clicking on the "Map Mode" menu. Options include terrain, topographic, satellite, street, National Geographic, and outline. The Outline mode allows you to click on a specific country and access a separate outline map of that country with features that you can turn on or off.
4. Click on the THEMES tab to display categories of data layers that can be overlaid on the map. Availability of these data layers varies depending on the zoom level of the map. Each data layer also has a legend and a transparency control.
5. Click on the DRAWING TOOLS tab to display a variety of tools that you can use to draw on the map. Most tools allow you to adjust the outline and fill colors, line width, and transparency. The tab can also be selected by clicking on the "Draw" icon.
6. Click on the MARKERS tab to display a variety of markers that you can drag and drop on the map. You can set the markers at three different sizes. The MARKERS tab can also be selected by clicking on the "Markers" icon.
7. To easily measure distances on the map, click on the "Measure" (ruler) icon. Click once on the map to start measuring, move the pointer to another spot, and click again to stop. The resulting line will display the distance in either kilometers or miles based on your selection from the dropdown menu.

COOPERATIVE LEARNING

Cooperative learning strategies transform today's classroom diversity into a vital resource for promoting students' acquisition of both challenging academic content and language. These strategies promote active engagement and social motivation for all students.

STRUCTURE & GRAPHIC	DESCRIPTION	BENEFITS & PURPOSES
CORNERS A strongly agree B disagree C agree D strongly disagree	• Corners of the classroom are designated for focused discussion of four aspects of a topic. • Students individually think and write about the topic for a short time. • Students group into the corner of their choice and discuss the topic. • At least one student from each corner shares about the corner discussion.	• By "voting" with their feet, students literally take a position about a topic. • Focused discussion develops deeper thought about a topic. • Students experience many valid points of view about a topic.
FISHBOWL 1 2 3 4 5 6 7 8 9 10	• Part of the class sits in a close circle facing inward; the other part of the class sits in a larger circle around them. • Students on the inside discuss a topic while those outside listen for new information and/or evaluate the discussion according to pre-established criteria. • Groups reverse positions.	• Focused listening enhances knowledge acquisition and listening skills. • Peer evaluation supports development of specific discussion skills. • Identification of criteria for evaluation promotes self-monitoring.
INSIDE-OUTSIDE CIRCLE A B C D E F G H I J 1 2 3 4 5 6 7 8 9 10	• Students stand in concentric circles facing each other. • Students in the outside circle ask questions; those inside answer. • On a signal, students rotate to create new partnerships. • On another signal, students trade inside/outside roles.	• Talking one-on-one with a variety of partners gives risk-free practice in speaking skills. • Interactions can be structured to focus on specific speaking skills. • Students practice both speaking and active listening.
JIGSAW Expert Group 1 (A B C D) — A's Expert Group 2 (A B C D) — B's Expert Group 3 (A B C D) — C's Expert Group 4 (A B C D) — D's	• Students are grouped evenly into "expert" groups. • Expert groups study one topic or aspect of a topic in depth. • Students regroup so that each new group has at least one member from each expert group. • Experts report on their study. Other students learn from the experts.	• Becoming an expert provides in-depth understanding in one aspect of study. • Learning from peers provides breadth of understanding of over-arching concepts.

STRUCTURE & GRAPHIC	DESCRIPTION	BENEFITS & PURPOSES
NUMBERED HEADS Think Time Talk Time Share 2's Time	• Students number off within each group. • Teacher prompts or gives a directive. • Students think individually about the topic. • Groups discuss the topic so that any member of the group can report for the group. • Teacher calls a number and the student with that number reports for the group.	• Group discussion of topics provides each student with language and concept understanding. • Random recitation provides an opportunity for evaluation of both individual and group progress.
ROUNDTABLE	• Students sit around tables in groups of four. • Teacher asks a question with many possible answers. • Each student around the table answers the question a different way.	• Encouraging elaboration creates appreciation for diversity of opinion and thought. • Eliciting multiple answers enhances language fluency.
TEAM WORD WEBBING	• Teams of students sit around a large piece of paper. Each team member has a different colored marker. • Teacher assigns a topic for a Word Web. • Each student adds to the part of the web nearest to him/her. • On a signal, students rotate the paper and each student adds to the nearest part again.	• Individual input to a group product ensures participation by all students. • Shifting point of view supports both broad and in-depth understanding of concepts.
THINK, PAIR, SHARE Think Pair Share	• Students think about a topic suggested by the teacher. • Pairs discuss the topic. • Students individually share information with the class.	• The opportunity for self-talk during the individual think time allows the student to formulate thoughts before speaking. • Discussion with a partner reduces performance anxiety and enhances understanding.
THREE-STEP INTERVIEW GROUP	• Students form pairs. • Student A interviews Student B about a topic. • Partners reverse roles. • Student A shares with the class information from Student B; then B shares information from Student A.	• Interviewing supports language acquisition by providing scripts for expression. • Responding provides opportunities for structured self-expression.

UNIT 11

Southeast Asia

TECHTREK

myNGconnect.com

Floating market in Damnern Saduak, Thailand

Digital Library
Unit 11 GeoVideo
Introduce Southeast Asia

Explorer Video Clip
Kristofer Helgen, Zoologist
National Geographic Emerging Explorer

NATIONAL GEOGRAPHIC **PHOTO GALLERY**
Regional photos, including the Mekong and Irrawaddy rivers, Angkor Wat, and Hanoi and Jakarta

Maps and Graphs
Interactive Map Tool

Interactive Whiteboard GeoActivities
- Analyze Fishing on the Mekong River
- Research the Borobudur Temple
- Map Southeast Asia's Spice Trade

Connect to NG
Research links and current events in Southeast Asia

NATIONAL GEOGRAPHIC

ATLAS

Page numbers of maps match Student Edition.
For a full Atlas go to the Teacher's Reference Guide.

World Physical

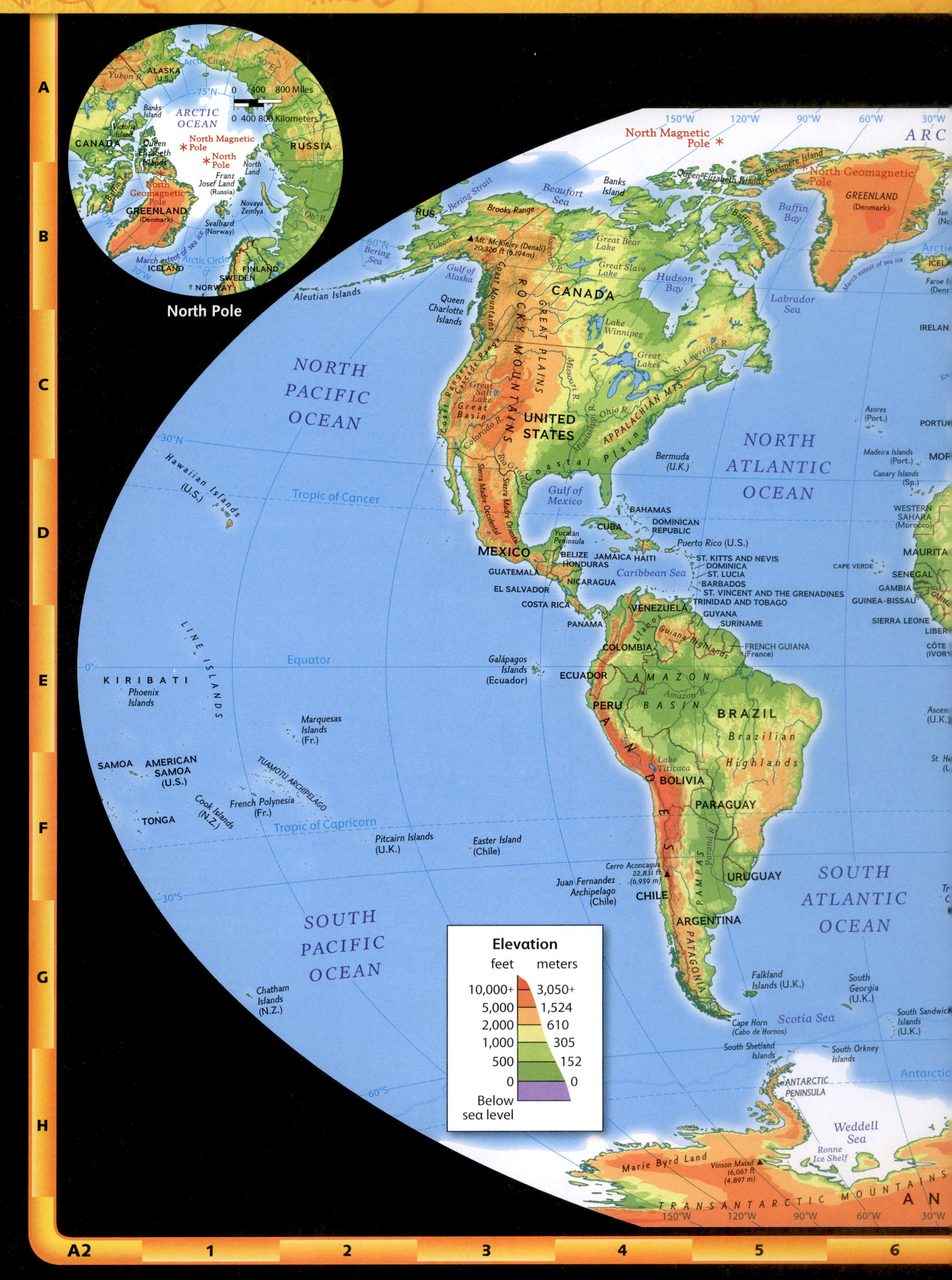

ANTARCTICA
QUEEN MAUD LAND
Weddell Sea
Antarctic Peninsula
Ronne Ice Shelf
Vinson Massif 16,067 ft (4,897 m)
WEST ANTARCTICA
South Pole
South Geomagnetic Pole
EAST ANTARCTICA
Ross Ice Shelf
Ross Sea
South Magnetic Pole
WILKES LAND
ATLANTIC OCEAN
INDIAN OCEAN
PACIFIC OCEAN
Antarctic Circle
September extent of sea ice
0 400 800 Miles
0 400 800 Kilometers
South Pole
Severnaya Zemlya
New Siberian Islands
East Siberian Sea
Wrangel I.
Franz Josef Land
Barents Sea
Novaya Zemlya
Kara Sea
Laptev Sea
Central Siberian Plateau
West Siberian Plain
RUSSIA
Bering Sea
Sea of Okhotsk
Aleutian Islands
Lake Baikal
KAZAKHSTAN
Kazakh Uplands
MONGOLIA
Mongolia Plateau
GOBI
Sea of Japan (East Sea)
NORTH KOREA
SOUTH KOREA
JAPAN
Caspian Sea
Aral Sea
Black Sea
TURKEY
UZBEKISTAN
TURKMENISTAN
KYRGYZSTAN
TAJIKISTAN
AFGHANISTAN
Taklimakan Desert
Kunlun Mountains
Plateau of Tibet
CHINA
HIMALAYA
Mt. Everest 29,035 ft (8,850 m)
NEPAL
BHUTAN
PAKISTAN
IRAN
IRAQ
SYRIA
CYPRUS
LEBANON
ISRAEL
JORDAN
KUWAIT
SAUDI ARABIA
ARABIAN PENINSULA
QATAR
OMAN
YEMEN
Zagros Mountains
INDIA
Deccan Plateau
Arabian Sea
Bay of Bengal
SRI LANKA
MALDIVES
East China Sea
TAIWAN
PHILIPPINES
South China Sea
VIETNAM
THAILAND
LAOS
CAMB.
MALAYSIA
SINGAPORE
BRUNEI
Borneo
INDONESIA
Sumatra
Java
Java Sea
Celebes
Celebes Sea
Banda Sea
New Guinea
PAPUA NEW GUINEA
TIMOR-LESTE (EAST TIMOR)
Timor
Timor Sea
Arafura Sea
Philippine Sea
Bonin Islands (Japan)
Volcano Islands (Japan)
Izu Islands (Japan)
Minami Tori Shima (Marcus) (Japan)
NORTH PACIFIC OCEAN
Northern Mariana Islands (U.S.)
Guam (U.S.)
PALAU
FEDERATED STATES OF MICRONESIA
MARSHALL ISLANDS
NAURU
KIRIBATI
SOLOMON ISLANDS
TUVALU
VANUATU
FIJI
Coral Sea
New Caledonia (Fr.)
Norfolk Island (Aus.)
AUSTRALIA
Great Sandy Desert
Western Plateau
Great Victoria Desert
Great Australian Bight
Great Artesian Basin
Great Barrier Reef
GREAT DIVIDING RANGE
Darling R.
Mt. Kosciuszko (2,228 m) 7,310 ft
Tasmania
Tasman Sea
NEW ZEALAND
North Island
South Island
Auckland Islands (N.Z.)
INDIAN OCEAN
Andaman Islands (India)
Nicobar Islands (India)
Andaman Sea
Chagos Archipelago (U.K.)
Diego Garcia (U.K.)
Cocos (Keeling) Islands (Aus.)
Christmas Island (Aus.)
Socotra (Yemen)
SEYCHELLES
COMOROS
MAURITIUS
Réunion (Fr.)
Île Amsterdam (Fr.)
Crozet Islands (Fr.)
Prince Edward Islands (South Africa)
Kerguelen Islands (Fr.)
Heard Island and McDonald Islands (Aus.)
September extent of sea ice
LIBYA
EGYPT
CHAD
SUDAN
SOUTH SUDAN
ERITREA
DJIBOUTI
ETHIOPIA
SOMALIA
CENTRAL AFRICAN REP.
UGANDA
KENYA
RWANDA
BURUNDI
DEM. REP. OF THE CONGO
CONGO
Congo Basin
Kilimanjaro 19,340 ft (5,895 m)
Lake Victoria
Lake Tanganyika
TANZANIA
ANGOLA
ZAMBIA
MALAWI
MOZAMBIQUE
ZIMBABWE
NAMIBIA
BOTSWANA
KALAHARI DESERT
Mozambique Channel
MADAGASCAR
SWAZILAND
LESOTHO
SOUTH AFRICA
Cape of Good Hope
Great Rift Valley
Red Sea
Gulf of Aden
Persian Gulf
Gulf of Oman
Elbrus 18,510 ft (5,642 m)
GEORGIA
UKRAINE
BELARUS
MOLDOVA
ESTONIA
LATVIA
POLAND
FINLAND
Kola Pen.
Ural Mountains
Volga R.
Ob R.
Lena R.
Yenisey R.
Altay Mountains
Tian Shan
Ganges R.
Yangtze R.
Yellow R.
Kuril Islands
Sakhalin
Kamchatka Peninsula
Hainan
Hokkaido
South Magnetic Pole
South Geomagnetic Pole
WILKES LAND
QUEEN MAUD LAND
0 1,000 2,000 Miles
0 1,000 2,000 Kilometers
60°E
90°E
120°E
150°E
30°N
0°
30°S
60°S
A
B
C
D
E
F
G
H
7
8
9
10
11
12

World Political

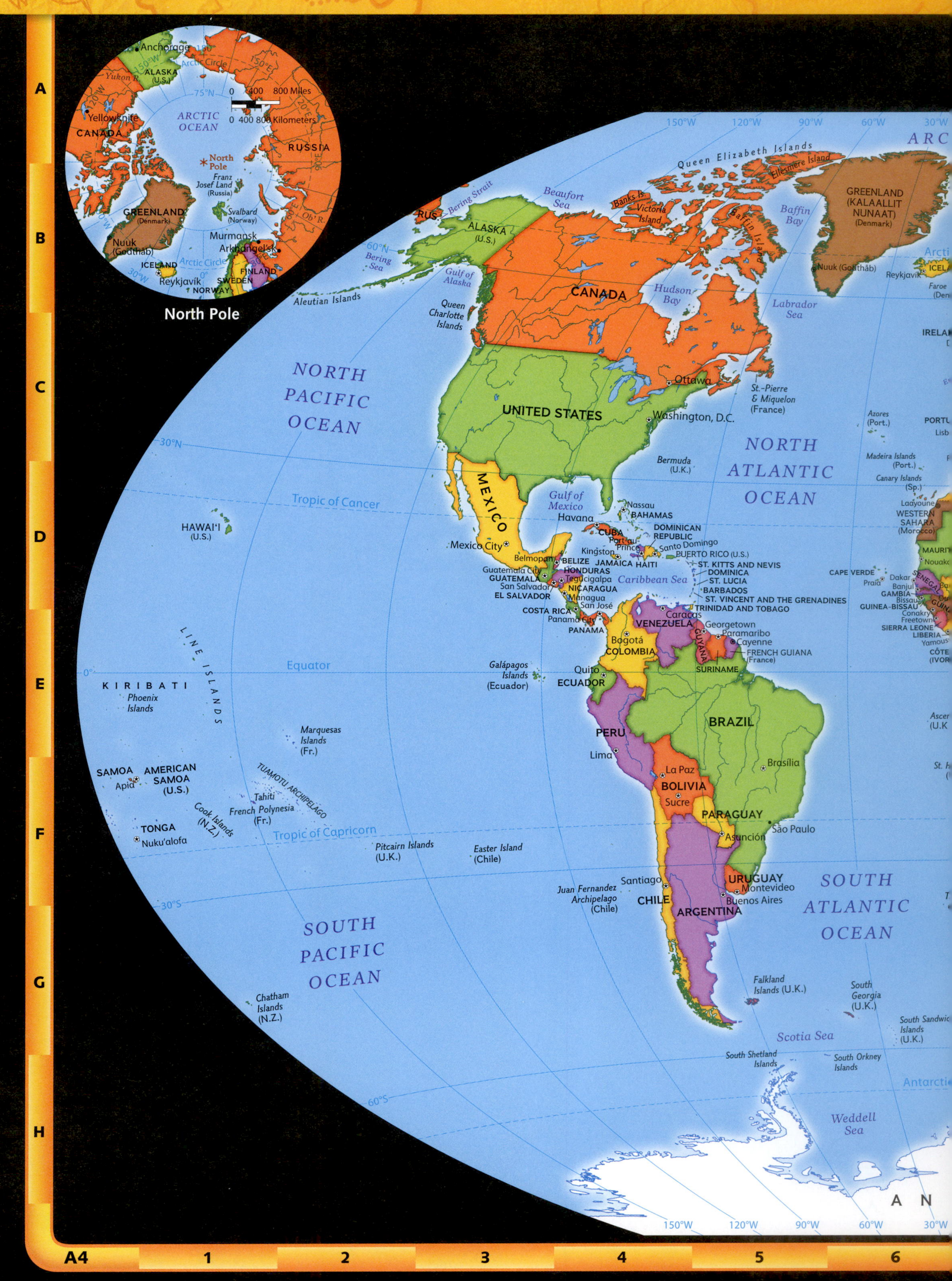

ANTARCTICA
QUEEN MAUD LAND
WILKES LAND
Antarctic Peninsula
Weddell Sea
Ronne Ice Shelf
Ross Ice Shelf
Ross Sea
Vinson Massif 16,067 ft (4,897 m)
South Pole
ATLANTIC OCEAN
INDIAN OCEAN
PACIFIC OCEAN
Antarctic Circle
0 400 800 Miles
0 400 800 Kilometers

South Pole

RUSSIA
KAZAKHSTAN
MONGOLIA
CHINA
JAPAN
INDIA
IRAN
AUSTRALIA
NEW ZEALAND
PAPUA NEW GUINEA
PHILIPPINES
INDONESIA
MALAYSIA
NORTH PACIFIC OCEAN
INDIAN OCEAN
Arabian Sea
Bay of Bengal
South China Sea
Philippine Sea
Coral Sea
Tasman Sea
FEDERATED STATES OF MICRONESIA
MARSHALL ISLANDS
SOLOMON ISLANDS
KIRIBATI
TUVALU
NAURU
PALAU
VANUATU
FIJI
MADAGASCAR
SOUTH AFRICA
SUDAN
SOUTH SUDAN
CHAD
LIBYA
EGYPT
SAUDI ARABIA
ETHIOPIA
SOMALIA
KENYA
TANZANIA
ANGOLA
ZAMBIA
NAMIBIA
BOTSWANA

0 1,000 2,000 Miles
0 1,000 2,000 Kilometers

A B C D E F G H
7 8 9 10 11 12 A5

Southeast Asia Physical

Southeast Asia Political
NATIONAL GEOGRAPHIC
CHINA
INDIA
MYANMAR (BURMA)
LAOS
THAILAND
VIETNAM
CAMBODIA
INDOCHINA PENINSULA
MALAY PENINSULA
MALAYSIA
SINGAPORE
BRUNEI
PHILIPPINES
INDONESIA
TIMOR-LESTE (EAST TIMOR)
SOUTH KOREA
JAPAN
TAIWAN
AUSTRALIA
PACIFIC OCEAN
INDIAN OCEAN
South China Sea
Philippine Sea
Andaman Sea
Gulf of Thailand
Sulu Sea
Celebes Sea
Java Sea
Banda Sea
Flores Sea
Timor Sea
Arafura Sea
Tropic of Cancer
Equator
Tropic of Capricorn
Nay Pyi Taw
Yangon (Rangoon)
Hanoi
Vientiane
Krung Thep (Bangkok)
Phnom Penh
Da Nang
Ho Chi Minh City (Saigon)
Manila
Cagayan de Oro
Davao
Bandar Seri Begawan
Kuala Lumpur
Banda Aceh
Medan
Jambi
Palembang
Jakarta
Bandung
Semarang
Surabaya
Balikpapan
Ujungpandang (Makassar)
Manado
Ambon
Dili
Jayapura
LUZON
Mindoro
Palawan
MINDANAO
BORNEO
SUMATRA
JAVA
CELEBES
MOLUCCAS
NEW GUINEA
Bali
Sumbawa
Sumba
Flores
Timor
Red R.
Black R.
Salween R.
Irrawaddy R.
Ping R.
Mekong R.
Kapuas R.
100°E
110°E
120°E
130°E
140°E
30°N
20°N
10°N
0°
10°S
20°S
30°S
N
S
E
W
0 300 600 Miles
0 300 600 Kilometers
A
B
C
D
E
F
G
H
1
2
3
4
5
6
A35

explore Southeast Asia with NATIONAL GEOGRAPHIC

MEET THE EXPLORER

NATIONAL GEOGRAPHIC

Emerging Explorer Jenny Daltry searches for unknown species of snakes, frogs, and crocodiles in unexplored corners of South Asia and Southeast Asia. Her work helps conserve these animals' habitats. Here, she inspects the fangs of a snake.

INVESTIGATE GEOGRAPHY

Mount Merapi is a volcanic mountain peak located near the center of the densely populated island of Java, Indonesia. It is the most active of the country's volcanoes. Its ash creates fertile soil, luring farmers in spite of the dangers.

STEP INTO HISTORY

A Buddhist monk prays at a statue in the Angkor Wat temple complex, Cambodia. Angkor Wat was originally built as a Hindu worship center in the 12th century. It remains the largest religious structure in the world.

594 UNIT 11

ONLINE WORLD ATLAS

Go to myNGconnect.com for maps of Southeast Asia.

CONNECT WITH THE CULTURE

The busy floating market in Damnern Saduak, Thailand has been attracting buyers since 1872. Hats, shown here, along with fruits, vegetables, flowers, and other food, are available.

595

CHAPTER PLANNER

SECTION SUPPORT

TE Resource Bank

myNGconnect.com

SECTION 1 GEOGRAPHY

1.1 Physical Geography

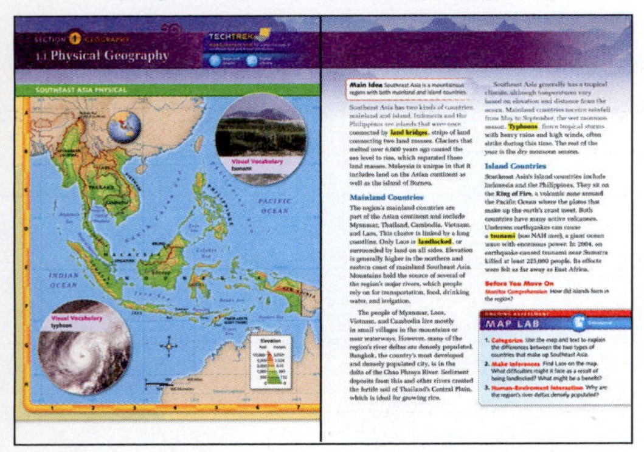

OBJECTIVE Analyze the location and physical geography of Southeast Asia.

Reading and Note-Taking
Answer Questions

Vocabulary Practice
Comparison Paragraphs

Whiteboard Ready!

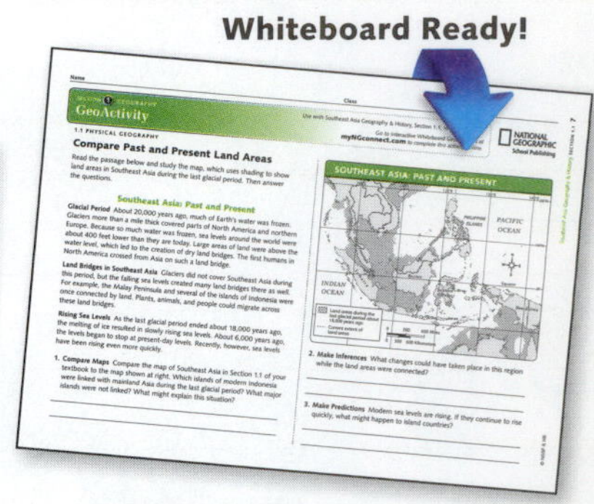

GeoActivity
Compare Past and Present Land Areas

SECTION 1 GEOGRAPHY

1.2 Parallel Rivers

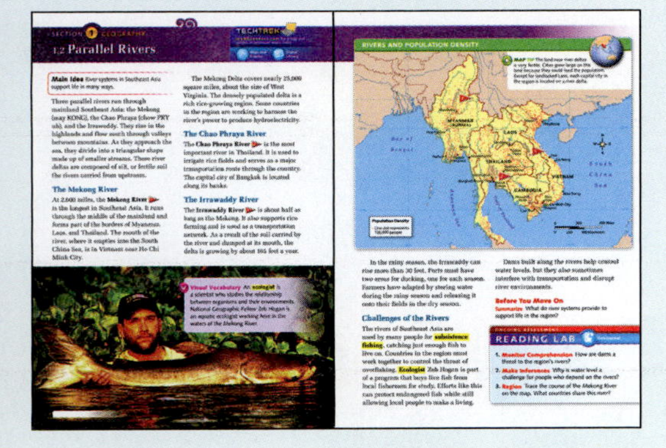

OBJECTIVE Understand the importance of Southeast Asia's major rivers for sustaining life in the region.

Reading and Note-Taking
Organize Information

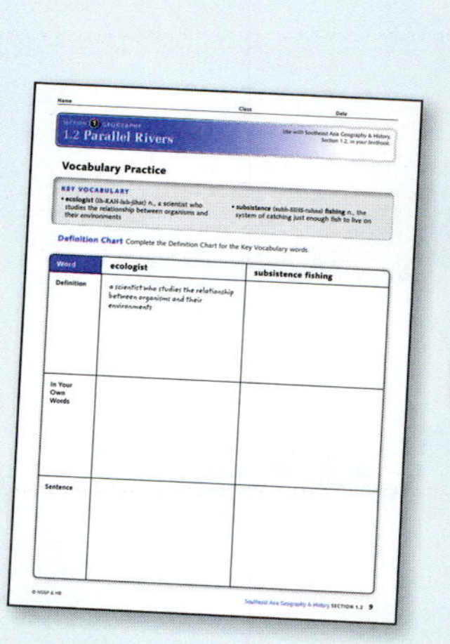

Vocabulary Practice
Definition Chart

Whiteboard Ready!

GeoActivity
Research an Environmental Issue

SECTION 1 GEOGRAPHY

1.3 The Malay Peninsula

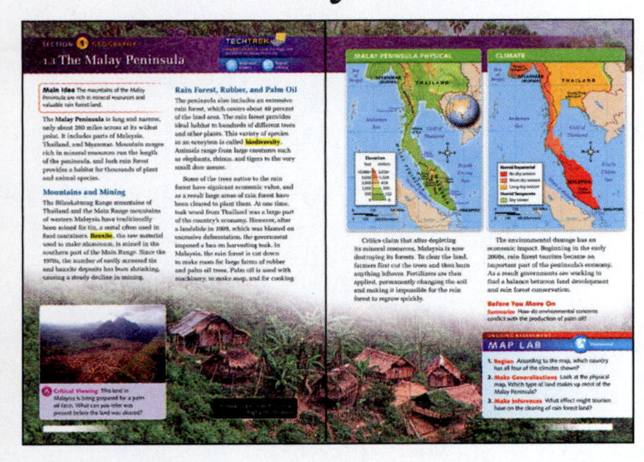

OBJECTIVE Examine the main features that characterize the Malay Peninsula.

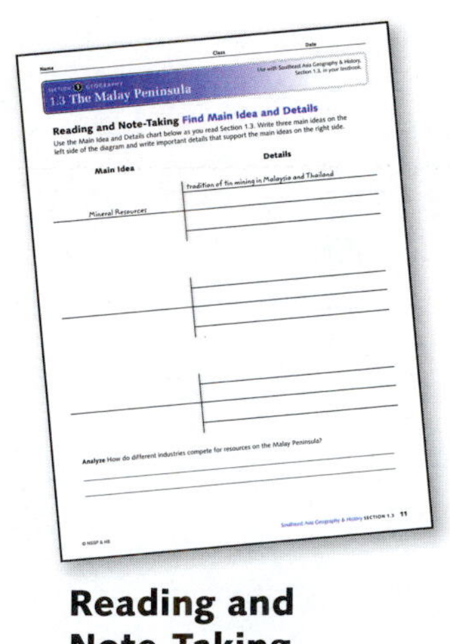

Reading and Note-Taking
Find the Main Idea and Details

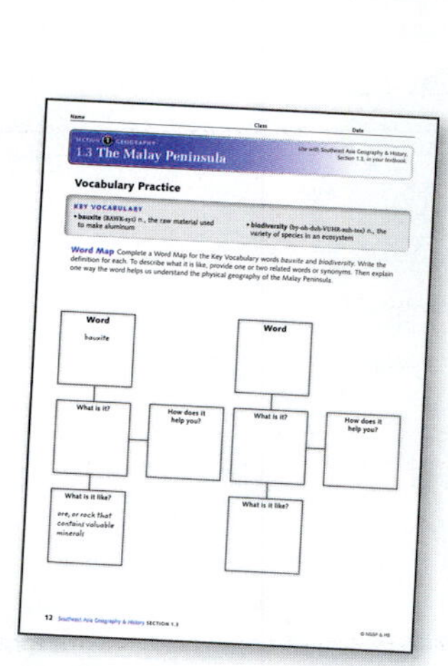

Vocabulary Practice
Word Map

Whiteboard Ready!

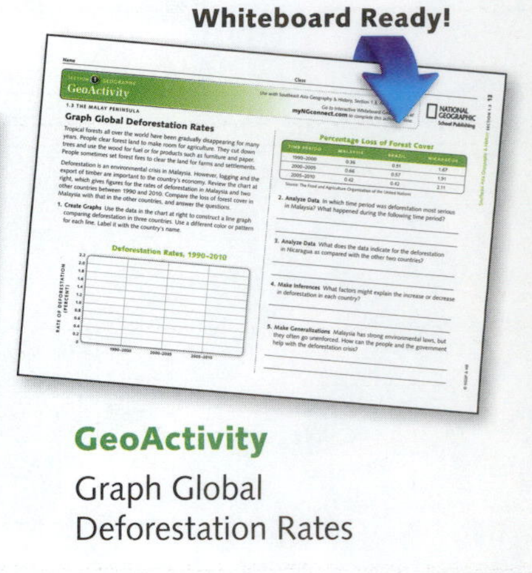

GeoActivity
Graph Global Deforestation Rates

ASSESSMENT

Student Edition
Ongoing Assessment: Map Lab

Resource Bank and myNGconnect.com
Review and Assessment, Sections 1.1–1.5

ExamView®
Test Generator CD-ROM
Section 1 Quiz in English and Spanish

Student Edition
Ongoing Assessment: Reading Lab

Resource Bank and myNGconnect.com
Review and Assessment, Sections 1.1–1.5

ExamView®
Test Generator CD-ROM
Section 1 Quiz in English and Spanish

Student Edition
Ongoing Assessment: Map Lab

Resource Bank and myNGconnect.com
Review and Assessment, Sections 1.1–1.5

ExamView®
Test Generator CD-ROM
Section 1 Quiz in English and Spanish

TECHTREK myNGconnect.com

Core Content Presentations
Teach *Physical Geography*

Digital Library
GeoVideo: *Introduce Southeast Asia*

Maps and Graphs
- **Interactive Map Tool** Analyze Settlement Patterns
- Online World Atlas: Southeast Asia Physical

Connect to NG
Research Links

Also Check Out
- NG Photo Gallery in **Digital Library**
- GeoJournal in **Student eEdition**

Fast Forward!
Core Content Presentations
Teach *Parallel Rivers*

Digital Library
NG Photo Gallery, Section 1

Connect to NG
Research Links

Also Check Out
- Graphic Organizers in **Teacher Resources**
- Online World Atlas in **Maps and Graphs**
- GeoJournal in **Student eEdition**

Fast Forward!
Core Content Presentations
Teach *The Malay Peninsula*

Maps and Graphs
Online World Atlas: Malay Peninsula Physical; Climate

Digital Library
NG Photo Gallery, Section 1

Connect to NG
Research Links

Interactive Whiteboard
GeoActivity Graph Global Deforestation Rates

Also Check Out
GeoJournal in **Student eEdition**

CHAPTER PLANNER

SECTION SUPPORT

SECTION 1 GEOGRAPHY

1.4 Island Nations

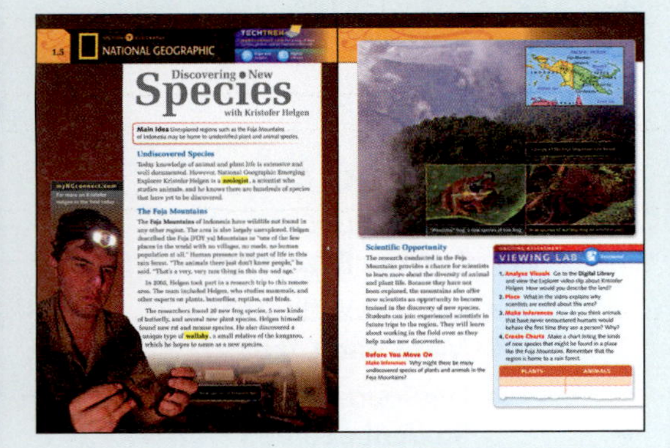

OBJECTIVE Draw conclusions about how geographic conditions affect life in the island nations.

Reading and Note-Taking
Outline and Take Notes

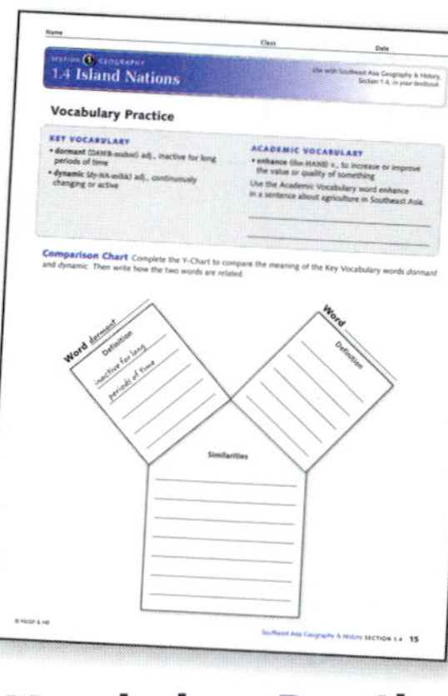

Vocabulary Practice
Comparison Chart

Whiteboard Ready!

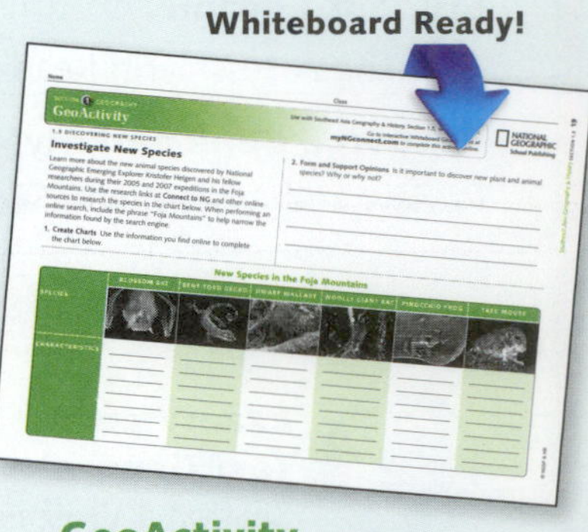

GeoActivity
Analyze the Effects of Krakatoa

SECTION 1 GEOGRAPHY

1.5 Discovering New Species

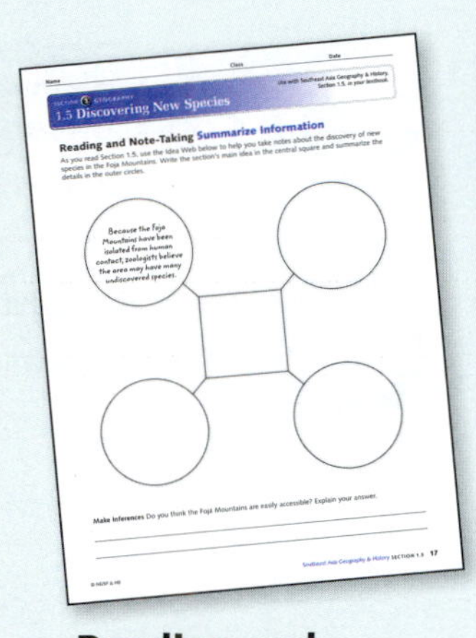

OBJECTIVE Analyze the importance of unexplored areas of Southeast Asia.

Reading and Note-Taking
Summarize Information

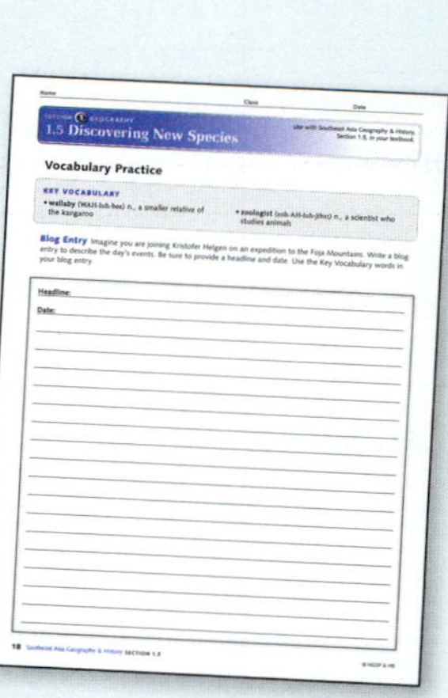

Vocabulary Practice
Blog Entry

Whiteboard Ready!

GeoActivity
Investigate New Species

SECTION 2 HISTORY

2.1 Ancient Valley Kingdoms

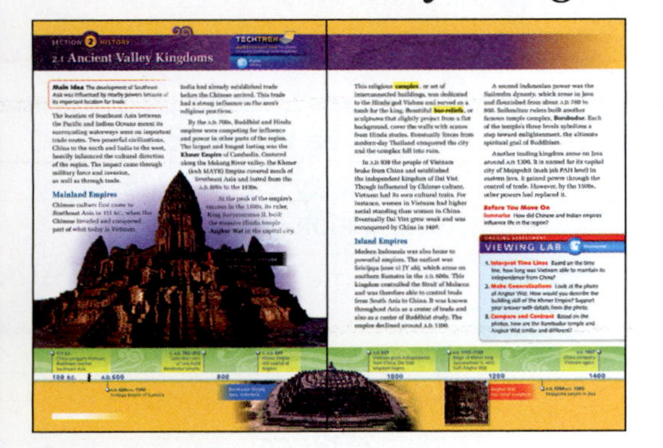

OBJECTIVE Analyze the role of physical geography in the history and culture of Southeast Asia.

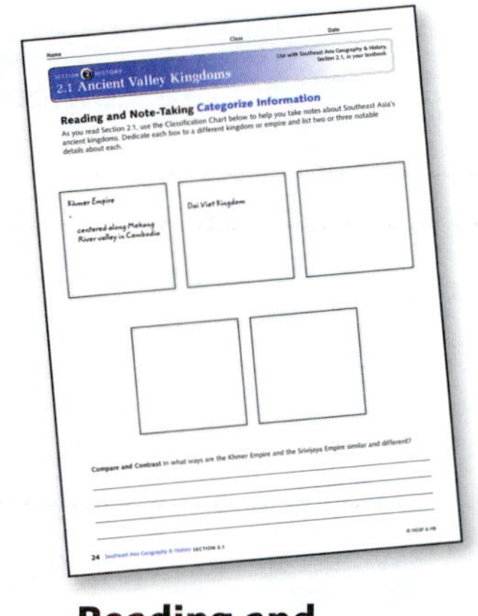

Reading and Note-Taking
Categorize Information

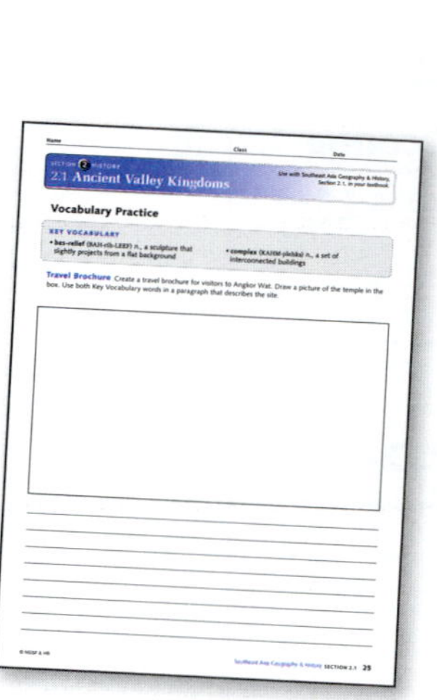

Vocabulary Practice
Travel Brochure

Whiteboard Ready!

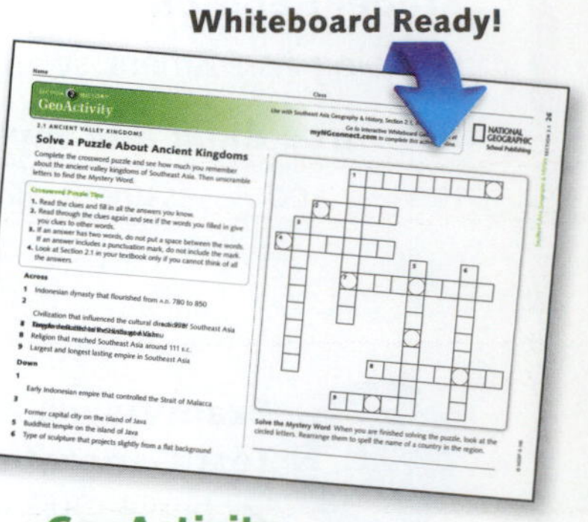

GeoActivity
Solve a Puzzle About Ancient Kingdoms

ASSESSMENT

Student Edition
Ongoing Assessment: Map Lab

Resource Bank and myNGconnect.com
Review and Assessment, Sections 1.1–1.5

ExamView®
Test Generator CD-ROM
Section 1 Quiz in English and Spanish

Student Edition
Ongoing Assessment: Viewing Lab

Teacher's Edition
Performance Assessment: Prepare a Museum Exhibit

Resource Bank and myNGconnect.com
Review and Assessment, Sections 1.1–1.5

ExamView®
Test Generator CD-ROM
Section 1 Quiz in English and Spanish

Student Edition
Ongoing Assessment: Viewing Lab

Resource Bank and myNGconnect.com
Review and Assessment, Sections 2.1–2.4

ExamView®
Test Generator CD-ROM
Section 2 Quiz in English and Spanish

TECHTREK myNGconnect.com

Fast Forward!
Core Content Presentations
Teach *Island Nations*

Interactive Whiteboard
GeoActivity Analyze the Effects of Krakatoa

Maps and Graphs
Online World Atlas: Tectonic Plates and Volcanoes; Population Density

Also Check Out
- NG Photo Gallery in **Digital Library**
- Graphic Organizers in **Teacher Resources**
- GeoJournal in **Student eEdition**

Fast Forward!
Core Content Presentations
Teach *Discovering New Species*

Interactive Whiteboard
GeoActivity Investigate New Species

Digital Library
Explorer Video Clip: *Kristofer Helgen*

Connect to NG
Research Links

Also Check Out
- NG Photo Gallery in **Digital Library**
- GeoJournal in **Student eEdition**

Fast Forward!
Core Content Presentations
Teach *Ancient Valley Kingdoms*

Maps and Graphs
Online World Atlas: Southeast Asia Political

Digital Library
GeoVideo: *Introduce Southeast Asia*

Connect to NG
Research Links

Also Check Out
- NG Photo Gallery in **Digital Library**
- Graphic Organizers in **Teacher Resources**
- GeoJournal in **Student eEdition**

CHAPTER PLANNER

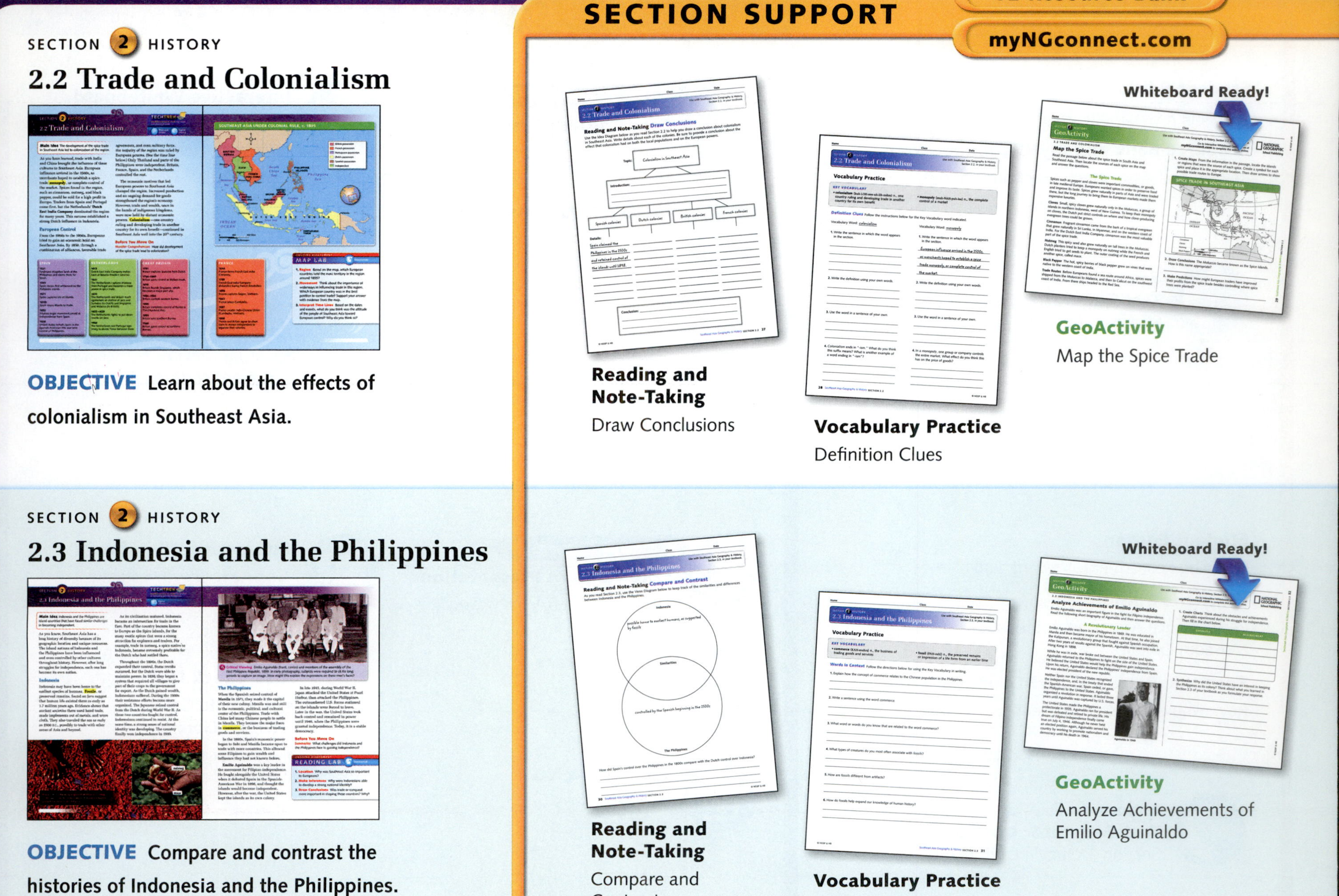

SECTION SUPPORT

TE Resource Bank

myNGconnect.com

SECTION 2 HISTORY

2.2 Trade and Colonialism

OBJECTIVE Learn about the effects of colonialism in Southeast Asia.

Reading and Note-Taking
Draw Conclusions

Vocabulary Practice
Definition Clues

Whiteboard Ready!

GeoActivity
Map the Spice Trade

SECTION 2 HISTORY

2.3 Indonesia and the Philippines

OBJECTIVE Compare and contrast the histories of Indonesia and the Philippines.

Reading and Note-Taking
Compare and Contrast

Vocabulary Practice
Words in Context

Whiteboard Ready!

GeoActivity
Analyze Achievements of Emilio Aguinaldo

ASSESSMENT

Student Edition
Ongoing Assessment: Map Lab

Resource Bank and myNGconnect.com
Review and Assessment, Sections 2.1–2.4

ExamView®
Test Generator CD-ROM
Section 2 Quiz in English and Spanish

Student Edition
Ongoing Assessment: Reading Lab

Resource Bank and myNGconnect.com
Review and Assessment, Sections 2.1–2.4

ExamView®
Test Generator CD-ROM
Section 2 Quiz in English and Spanish

TECHTREK myNGconnect.com

Fast Forward!
Core Content Presentations
Teach *Trade and Colonialism*

Connect to NG
Research Links

Digital Library
NG Photo Gallery, Section 2

Interactive Whiteboard
GeoActivity Map the Spice Trade

Maps and Graphs
Online World Atlas: Southeast Asia Under Colonial Rule, c. 1895

Also Check Out
GeoJournal in
Student eEdition

Fast Forward!
Core Content Presentations
Teach *Indonesia and the Philippines*

Digital Library
NG Photo Gallery, Section 2

Interactive Whiteboard
GeoActivity Analyze Achievements of Emilio Aguinaldo

Also Check Out
GeoJournal in
Student eEdition

CHAPTER PLANNER

SECTION 2 DOCUMENT-BASED QUESTION

2.4 The Vietnam War

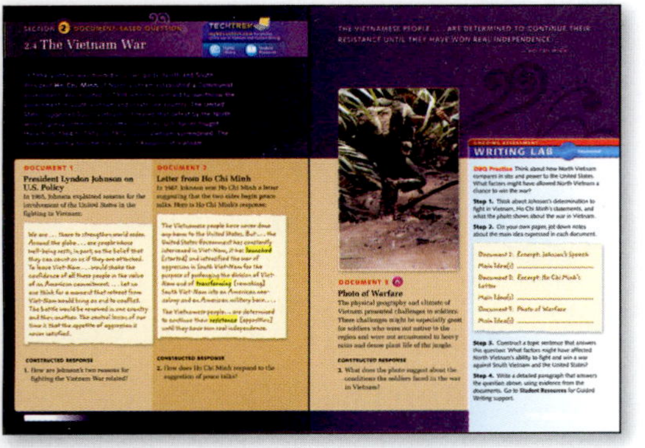

OBJECTIVE Use primary sources to explore the political reasons for the Vietnam War.

SECTION SUPPORT

TE Resource Bank

myNGconnect.com

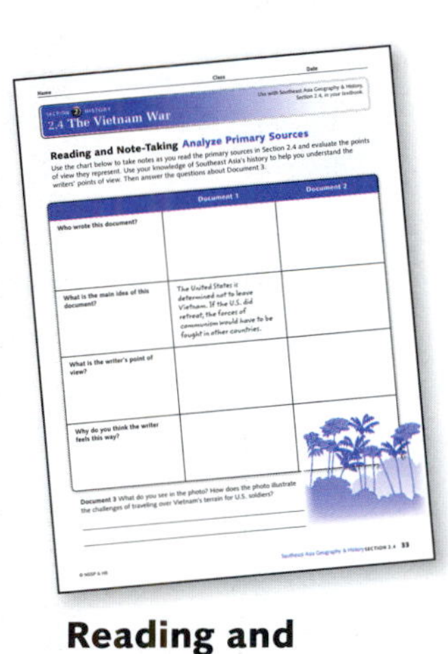

Reading and Note-Taking
Analyze Primary Sources

Vocabulary Practice
Word Squares

Whiteboard Ready!

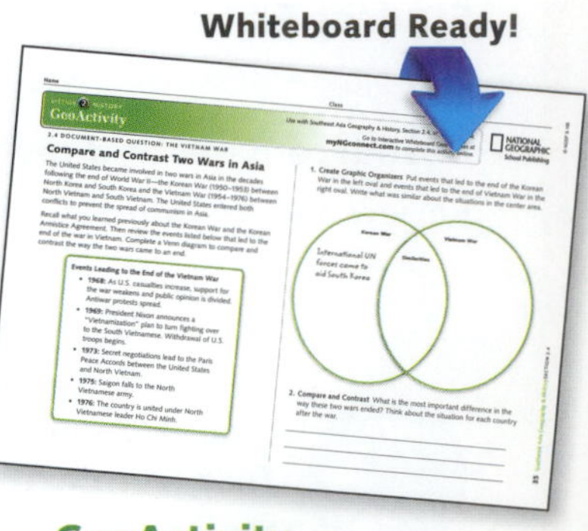

GeoActivity
Compare and Contrast Two Wars in Asia

CHAPTER ASSESSMENT

Review

INFORMAL ASSESSMENT

TE Resource Bank

myNGconnect.com

Review and Assessment

Standardized Test Practice

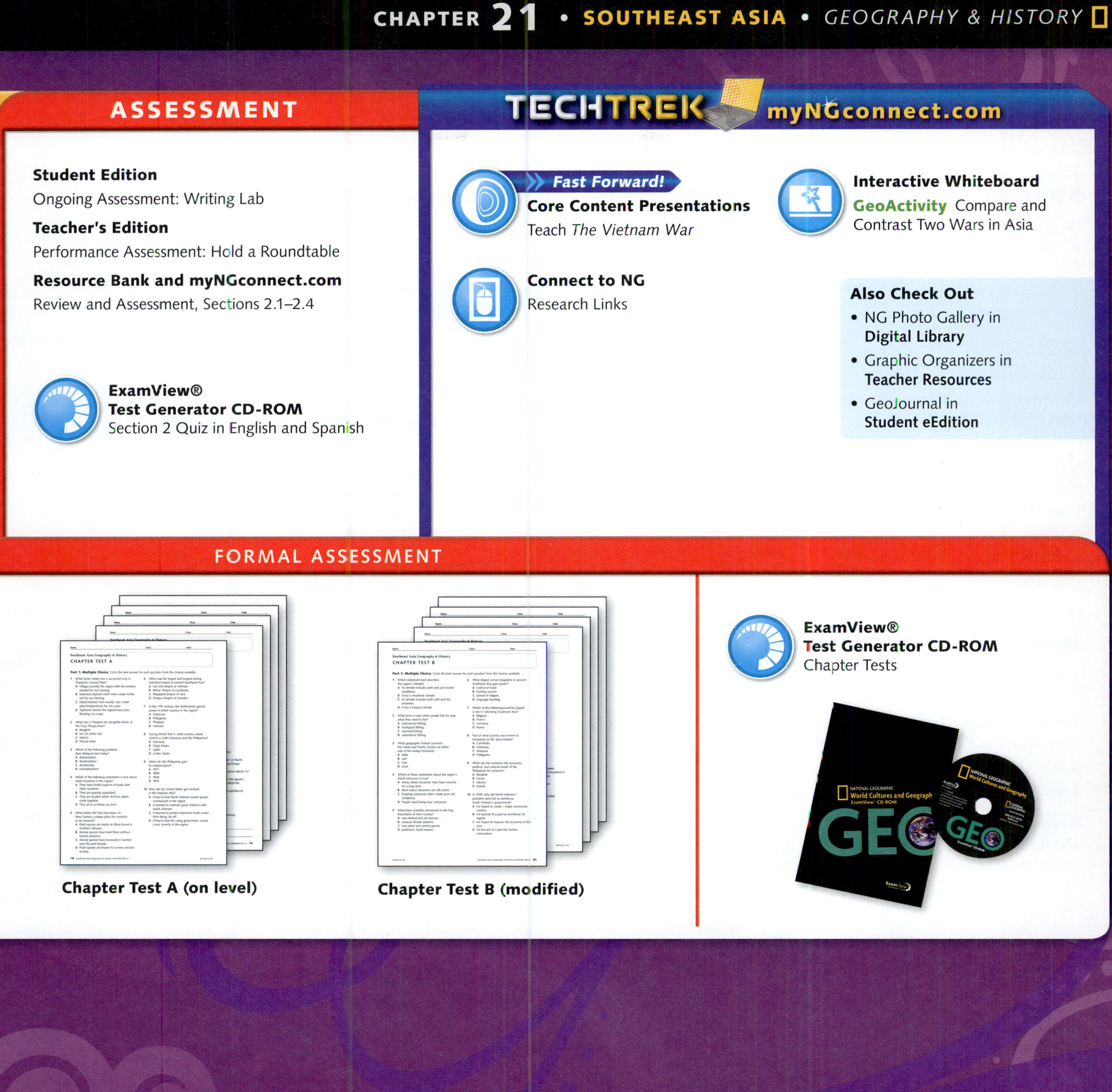

ASSESSMENT

Student Edition
Ongoing Assessment: Writing Lab

Teacher's Edition
Performance Assessment: Hold a Roundtable

Resource Bank and myNGconnect.com
Review and Assessment, Sections 2.1–2.4

ExamView®
Test Generator CD-ROM
Section 2 Quiz in English and Spanish

TECHTREK myNGconnect.com

Fast Forward!
Core Content Presentations
Teach *The Vietnam War*

Interactive Whiteboard
GeoActivity Compare and Contrast Two Wars in Asia

Connect to NG
Research Links

Also Check Out
- NG Photo Gallery in **Digital Library**
- Graphic Organizers in **Teacher Resources**
- GeoJournal in **Student eEdition**

FORMAL ASSESSMENT

Chapter Test A (on level)

Chapter Test B (modified)

ExamView®
Test Generator CD-ROM
Chapter Tests

STRATEGIES FOR DIFFERENTIATION

STRIVING READERS

eEdition Audiobook

Strategy 1 • Complete a 2 + 2 + 2 Chart

Display this chart and ask students to read and complete two examples for each category.

SOUTHEAST ASIA	
2 Kinds of Countries	• •
2 Important Physical Features	• •
2 Causes of Destruction	• •

Use with Section 1.1

Strategy 2 • Write a "Tweet"

Direct students to read and write a "Tweet" in their own words that explains the main idea of each paragraph. Have pairs of students compare their "Tweets" and agree on a final version.

Use with Sections 1.2–1.4

Strategy 3 • Ask Questions

1. Pairs read each section of the text and formulate one question that will help them understand it.
2. Pair One begins by asking Pair Two their question about the first section. Pair Two answers the question.
3. Pair One confirms the answer.
4. Pair Two chooses Pair Three to ask their question, and so on.

Use with All Sections *For Section 1.1, have students ask questions about mainland countries and island countries. For Section 2.4, have students ask questions about each document.*

Strategy 4 • Take Notes

Provide a copy of this T-Chart to students. Have them use it as they read to take notes about the two types of kingdoms. Tell them to add rows as needed.

Southeast Asian Empires

Mainland	Island

Use with Section 2.1 *You may wish to use this activity with Section 1.1, as a follow-up to Strategy 3.*

Strategy 5 • Do Pair Jigsaw Reading

Choose two lessons to assign to a pair of students. Each student will read and take notes on one lesson and report on it to the other student. Student presenters can also write 5Ws questions for the listeners to answer. You may wish to provide sample questions such as the following (for Section 2.1):

- **Who** was the temple at Angkor Wat dedicated to? *(Hindu god Vishnu)*
- **What** two civilizations strongly influenced Southeast Asia? *(China and India)*
- **When** did the independent kingdom of Dai Viet begin and end? *(939 and 1407)*
- **Where** was the Khmer Empire located? *(Cambodia, along the Mekong River valley)*
- **Why** was Southeast Asia's location important for trade routes? *(Its waterways run between the Pacific and Indian Oceans.)*

Use with Sections 2.1–2.3 *This activity can also be used with a larger group of students. For example, you may divide the class into five groups and assign each one a lesson from Section 1.*

INCLUSION

eEdition Audiobook

Strategy 1 • Clarify Countries

Provide an index card and instruct students to label one side "Mainland Countries" and the other side "Island Countries." Have them look at the map of the region and trace the outline of each country with their index finger. Then guide them in writing the name of each country on the correct side of the card, with space between each. (Note that Malaysia will appear on both sides.) As students read Section 1, have them trace significant physical features on each map and add details to the card about each country's geography.

Use with All Sections *Students may continue adding to their cards as they read Section 2, using time lines rather than maps.*

Strategy 2 • Provide a Chronological Summary

Give students copies of this chronology of major events around the Vietnam War and have them complete it as they read:

1800s	• ______ *(France)* controls Indochina, including Vietnam
______ *(1954)*	• France driven out • Vietnam divided into ______ *(North Vietnam)* and ______ *(South Vietnam)*
1964–73	• Vietnam War • Communist ______ *(North)* Vietnam tries to overthrow non-communist ______ *(South)* Vietnam • America fights on side of ______ *(South)* Vietnam
______ *(1975)*	• ______ *(North)* Vietnam defeats ______ *(South)* Vietnam • Country united as Vietnam, a Communist country

Use with Section 2.4

ENGLISH LANGUAGE LEARNERS

eEdition Audiobook Digital Library

Strategy 1 • Associate Word Pairs

Help students associate meaning with vocabulary words by giving them these sentences to complete with the words in parentheses. Have volunteers read completed sentences aloud.

1. (typhoon, hurricane) — A _______ *(typhoon)* is a fierce tropical storm similar to a _______ *(hurricane)*.
2. (tsunami, ocean) — A giant _______ *(ocean)* wave with enormous power is a _______ *(tsunami)*.

Use with Section 1.1 *Encourage native speakers or advanced ELL students to develop word pairs and sentences for beginning ELLs. Possible pairs might be* dormant/volcano, ecologist/zoologist, monopoly/control, *and* commerce/business.

Strategy 2 • Create Visual Vocabulary

Point out the Visual Vocabulary in Section 1.1 or project it from the **NG Photo Gallery**, and read aloud the definitions. Have students design their own Visual Vocabulary cards, starting with these two words and a simple sketch. As students read the chapter, suggest additional vocabulary words, such as *landlocked, land bridges, biodiversity,* and *dormant*.

Use with Section 1

Strategy 3 • Build a Word Family

Write the word *colony* on the board and ask a volunteer to use it in a sentence. Tell students that a variety of endings can be added to the word *colony* and that the new words that result all make up a word family built around the word *colony*. Then write the word equations below, one at a time. Have volunteers say the resulting word. When the list is completed, provide a brief definition for each word and allow students to work in pairs to use each word in a sentence.

colony + -ist = *colonist* = a person who makes a colony
colony + -ial = *colonial* = like a colony
colony + -ial + -ism = *colonialism* = the practice of making colonies
colony + -ize = *colonize* = to make a colony

Use with Section 2.2 *This activity can be extended for advanced English language learners with the word* monopoly *in Section 2.2 and the word* transform *in Section 2.4. Start with* monopolize *and* transformation. *Then suggest that students build words with the word parts* mono- *and* trans-.

GIFTED & TALENTED

Connect to NG Magazine Maker

Strategy 1 • Research and Report

Tell students to choose one country in the region and do research to report on at least six species of animals that are threatened with extinction in that country. Suggest that students include photographs or sketches of the animals along with details about them and possible threats to their existence.

Use with Sections 1.3 and 1.5

Strategy 2 • Explore History Through Art

Remind students that events in history impact the arts, and ask them to use the **Research Links** to find posters, photographs, and works of art on the subject of the Vietnam War. Suggest that they make a visual display with captions for their findings. Students may want to use the **Magazine Maker** to create their displays.

Use with Section 2.4

PRE-AP

Connect to NG

Strategy 1 • Compare Island Countries

Challenge students to come up with at least five categories to use to compare Indonesia and the Philippines. Suggest that they do research and design a Venn diagram or another kind of compare/contrast chart to include with their written comparison.

Use with Sections 1.4 and 2.3 *Students may extend their diagram or chart by adding Malaysia to their comparison.*

Strategy 2 • Infer the Impact of Colonialism

Ask students to consider what kind of evidence or traces of past control by a European country might still exist in any of the Southeast Asian countries. Suggest that they use the **Research Links** to find evidence to support their inferences. Have them work independently or in pairs to evaluate whether these influences have been positive or negative for the people of Southeast Asia. Students may express their ideas in a written or oral editorial.

Use with Sections 2.2 and 2.3

CHAPTER 21

Southeast Asia GEOGRAPHY & HISTORY

TECHTREK FOR THIS CHAPTER

Student eEdition · Maps and Graphs · Interactive Whiteboard GeoActivities · Digital Library · Connect to NG

Go to myNGconnect.com for more on Southeast Asia.

Komodo dragon

PREVIEW THE CHAPTER

Essential Question What are the geographic conditions that divide Southeast Asia into many different parts?

KEY VOCABULARY
- land bridge
- landlocked
- typhoon
- tsunami
- subsistence fishing
- ecologist
- bauxite
- biodiversity
- dynamic
- dormant
- zoologist
- wallaby

ACADEMIC VOCABULARY
enhance

TERMS & NAMES
- Ring of Fire
- Mekong River
- Chao Phraya River
- Irrawaddy River
- Malay Peninsula
- Foja Mountains

Essential Question How have physical barriers in Southeast Asia influenced its history?

KEY VOCABULARY
- complex
- bas-relief
- monopoly
- colonialism
- fossil
- commerce
- launch
- resistance

ACADEMIC VOCABULARY
transform

TERMS & NAMES
- Khmer Empire
- Angkor Wat
- Borobudur
- Dutch East India Company
- Manila
- Emilio Aguinaldo
- Ho Chi Minh

TECHTREK myNGconnect.com

Fast Forward!

Core Content Presentations
Introduce *Southeast Asia Geography & History*

Digital Library
- GeoVideo: *Introduce Southeast Asia*
- NG Photo Gallery

Maps and Graphs
- **Interactive Map Tool** Explore Southeast Asia's Barriers
- Online World Atlas: Southeast Asia Political

Also Check Out
Charts & Infographics and Graphic Organizers in **Teacher Resources**

INTRODUCE THE CHAPTER

INTRODUCE THE MAP

Use the Southeast Asia Political map to review the idea of a region. *(A region is an area that includes several countries with physical and cultural similarities.)*

Direct students' attention to the map. **ASK:** What physical characteristics link the countries in this region? *(Most countries are islands or have a border on the sea. Almost all areas in the region are within the tropics.)*

COMPARE ACROSS REGIONS

Download the Southeast Asia by the Numbers chart. **ASK:** How does Indonesia compare in area and population to the United States? *(Indonesia has about one-fifth the area of the United States but more than three-quarters the number of people.)* Point out that with such a high population and relatively small land area, Indonesia has a high population density. **ASK:** What issues might arise in areas with a high population density? *(crowded schools, lack of housing or jobs for everyone, overuse of resources, increased pollution)*

Southeast Asia BY THE NUMBERS

COUNTRY	LAND AREA (SQ MI)	POPULATION
Brunei	2,033	395,027
Cambodia	68,925	14,453,680
East Timor	5,743	1,154,625
Indonesia	699,449	242,968,342
Laos	89,112	6,368,162
Malaysia	126,895	28,274,729
Myanmar	252,320	53,414,374
Philippines	115,124	99,900,177
Singapore	265	4,701,069
Thailand	197,255	67,089,500
Vietnam	119,718	89,571,13
United States	3,537,454	307,212,123

Source: CIA World Factbook

INTRODUCE THE ESSENTIAL QUESTIONS

SECTION 1 • GEOGRAPHY

What are the geographic conditions that divide Southeast Asia into many different parts?

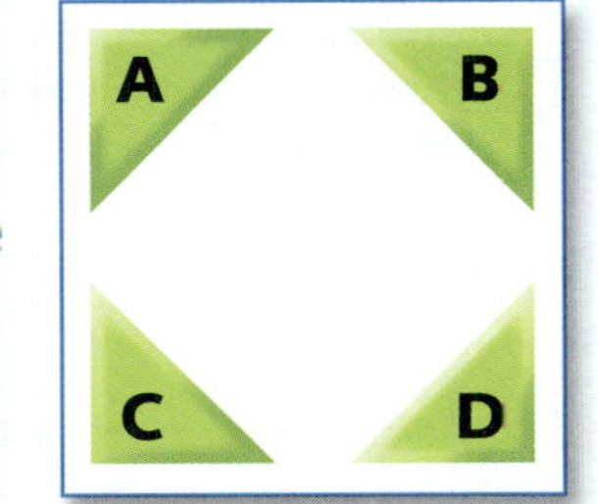

Four Corner Activity: The Impact of Being Divided This activity helps students explore the effect of a country being divided among more than one geographic unit. Post the four problems shown below in different corners of the classroom. Assign groups of students to each corner. Tell them that their task is to analyze one set of problems that might arise in a country that is divided into physically unconnected areas. **ASK:** Which is the worst problem that may result from being divided? Why would this problem be worse than the others? Discuss how a government might address the issues that a country with physical divisions might have to face. 0:25 minutes

A Cultural Differences The people of separate areas might have different cultures with separate ethnic groups, languages, religions, and histories. These differences can make it difficult to develop a sense of national unity.

B Uneven Economic Development Separate areas will have different quantities of resources and different levels of economic development. Tensions could arise between poorer and richer areas. Economic policies that seem to favor one area over another could be viewed with suspicion or anger.

C Separatist Political Movements Ethnic or other cultural differences might lead groups in one part of the country to push for independence. Such a movement would threaten the unity of the country. A harsh government response could lead to long-term conflict.

D Infrastructure Challenges A country with physical divisions faces unique challenges to building a strong infrastructure. Transportation between parts of the country could be difficult if the distances are great or if physical barriers block easy access. Health care and other services might be difficult to provide in parts of the country that are more remote. Communication systems would be more difficult to maintain and possibly unavailable in some areas. These infrastructure challenges can heighten cultural, political, or economic differences.

SECTION 2 • HISTORY

How have physical barriers in Southeast Asia influenced its history?

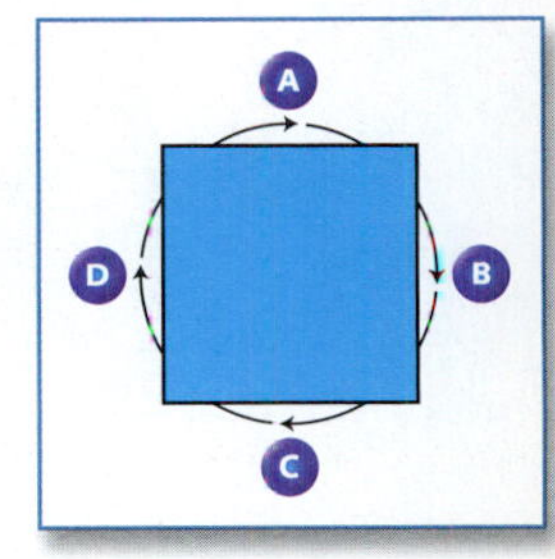

Team Word Webbing: Physical Barriers Have students form groups of four. Assign a letter from A to D to the students in each group. Tell students to open their text to the Southeast Asia Physical map. Beginning with Student A, have students take turns writing on a sheet of paper the physical barriers that they think might have affected the history of Southeast Asia. Remind them to look at the map to generate ideas. After the groups have had some time to add to their lists, call on volunteers, alternating groups, to state one of the barriers they identified. Discuss as a class which barrier might be the most significant. 0:25 minutes

ACTIVE OPTIONS

Interactive Map Tool

Explore Southeast Asia's Barriers

PURPOSE Explore physical barriers that influence life in Southeast Asia

SET-UP

1. Open the **Interactive Map Tool,** set the "Map Mode" to Topographic, and zoom in on the island of Borneo.
2. Under "Physical Systems—Land," turn on the Surface Elevations layer. Set the transparency level to about 35 percent.

ACTIVITY

Ask students to identify the parts of the island that they think might be less inhabited because of landforms. *(the mountains in the north)* Then click off the Surface Elevation layer. Under "Environment and Society," open the Land Cover layer with the transparency again set at about 35 percent. Call on students to name parts of the island that they think might not be inhabited because of heavy foresting. *(Most of the island seems to be covered by forests.)* Finally, turn off the Land Cover layer and, under "Human Systems—Populations & Culture," turn on the Population Density layer to check the predictions that students made. 0:20 minutes

INTRODUCE CHAPTER VOCABULARY

Teacher Resources

Word Mapping Divide the class into groups of six. Download the Word Map shown at right from the **Graphic Organizers** and distribute copies to students, or have them copy it onto their own paper. See "Best Practices for Active Teaching" for a review of the activity. Have each student in each group take three of the Key Vocabulary terms in the chapter and fill out the chart. Then have students teach their vocabulary words to other members of their group.

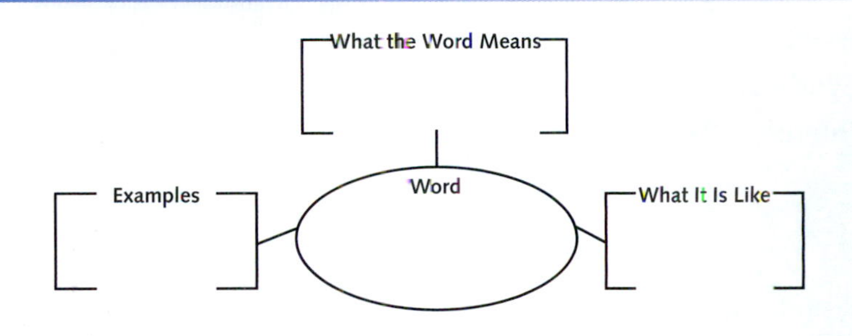

SECTION 1 GEOGRAPHY

1.1 Physical Geography

myNGconnect.com For a physical map of Southeast Asia and Visual Vocabulary

Maps and Graphs | Digital Library

SOUTHEAST ASIA PHYSICAL

Main Idea Southeast Asia is a mountainous region with both mainland and island countries.

Southeast Asia has two kinds of countries: mainland and island. Indonesia and the Philippines are islands that were once connected by **land bridges**, strips of land connecting two land masses. Glaciers that melted over 6,000 years ago caused the sea level to rise, which separated these land masses. Malaysia is unique in that it includes land on the Asian continent as well as the island of Borneo.

Mainland Countries

The region's mainland countries are part of the Asian continent and include Myanmar, Thailand, Cambodia, Vietnam, and Laos. This cluster is linked by a long coastline. Only Laos is **landlocked**, or surrounded by land on all sides. Elevation is generally higher in the northern and eastern coast of mainland Southeast Asia. Mountains hold the source of several of the region's major rivers, which people rely on for transportation, food, drinking water, and irrigation.

The people of Myanmar, Laos, Vietnam, and Cambodia live mostly in small villages in the mountains or near waterways. However, many of the region's river deltas are densely populated. Bangkok, the country's most developed and densely populated city, is in the delta of the Chao Phraya River. Sediment deposits from this and other rivers created the fertile soil of Thailand's Central Plain, which is ideal for growing rice.

Southeast Asia generally has a tropical climate, although temperatures vary based on elevation and distance from the ocean. Mainland countries receive rainfall from May to September, the wet monsoon season. **Typhoons**, fierce tropical storms with heavy rains and high winds, often strike during this time. The rest of the year is the dry monsoon season.

Island Countries

Southeast Asia's island countries include Indonesia and the Philippines. They sit on the **Ring of Fire**, a volcanic zone around the Pacific Ocean where the plates that make up the earth's crust meet. Both countries have many active volcanoes. Undersea earthquakes can cause a **tsunami** (soo NAH mee), a giant ocean wave with enormous power. In 2004, an earthquake-caused tsunami near Sumatra killed at least 225,000 people. Its effects were felt as far away as East Africa.

Before You Move On

Monitor Comprehension How did islands form in the region? Islands were formed when glaciers melted and the sea level rose, creating water barriers around land masses.

ONGOING ASSESSMENT

MAP LAB GeoJournal

1. **Categorize** Use the map and text to explain the differences between the two types of countries that make up Southeast Asia.
2. **Make Inferences** Find Laos on the map. What difficulties might it face as a result of being landlocked? What might be a benefit?
3. **Human-Enviroment Interaction** Why are the region's river deltas densely populated?

PLAN

OBJECTIVE Analyze the location and physical geography of Southeast Asia.

CRITICAL THINKING SKILLS FOR SECTION 1.1

- Main Idea
- Monitor Comprehension
- Categorize
- Make Inferences
- Synthesize
- Interpret Maps

PRINT RESOURCES

Teacher's Edition Resource Bank

- Reading and Note-Taking: Answer Questions
- Vocabulary Practice: Comparison Paragraphs
- **GeoActivity** Compare Past and Present Land Areas

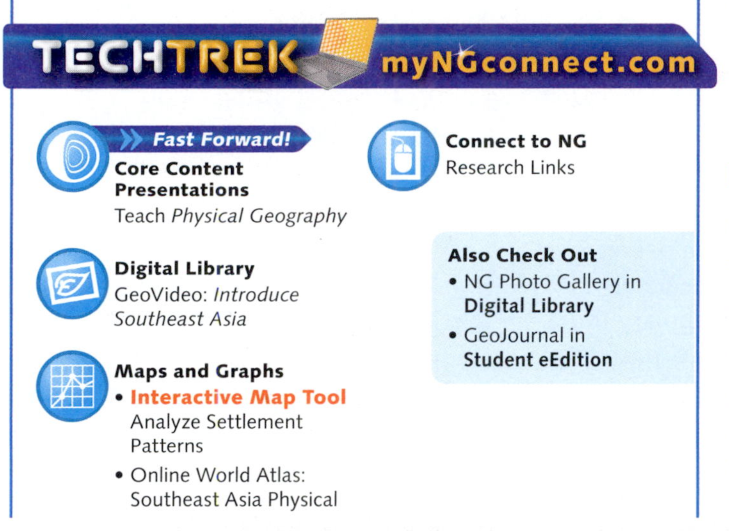

TECHTREK myNGconnect.com

Fast Forward!

Core Content Presentations
Teach *Physical Geography*

Digital Library
GeoVideo: *Introduce Southeast Asia*

Maps and Graphs
- **Interactive Map Tool** Analyze Settlement Patterns
- Online World Atlas: Southeast Asia Physical

Connect to NG
Research Links

Also Check Out
- NG Photo Gallery in **Digital Library**
- GeoJournal in **Student eEdition**

BACKGROUND FOR THE TEACHER

Two volcanic eruptions in the region have been of historic proportions. Krakatoa, in Indonesia, erupted in 1883. The explosion was so powerful that it could be heard more than 2,200 miles away. The eruption set off a tsunami that killed more than 35,000 people. In 1991, Mount Pinatubo in the Philippines had a spectacular eruption that shot so much dust into the atmosphere the climate around Earth was slightly cooled.

ESSENTIAL QUESTION

What are the geographic conditions that divide Southeast Asia into many different parts?

The mountains, seas, and rivers of the region create physical divisions within the countries of Southeast Asia. Section 1.1 explains these divisions and some of the region's environmental hazards.

INTRODUCE & ENGAGE

Digital Library

GeoVideo: *Introduce Southeast Asia* Show the section on physical geography to introduce students to the region and its physical characteristics. Then have students look at the Southeast Asia Physical map in the text. Call their attention to the country labels for Malaysia, Indonesia, and the Philippines. **ASK:** How are the borders of these countries different from those of Myanmar or Laos? *(They spread across many islands.)* Based on the video and the map, how do you think the physical characteristics of the region affect the lives of people who live there? *(People would need to adapt their farming and transportation to the numerous water systems in the region. People who live in island countries would need to adapt to being physically separated from parts of their own country.)* 0:15 minutes

TEACH

Maps and Graphs

Guided Discussion

1. **Synthesize** Direct students to the elevation legend on the map. **ASK:** Which mainland country is the most mountainous? *(Laos)* Which mainland country is mostly lowland? *(Cambodia)* How would you expect physical geography to affect settlement patterns in each country? *(People in Laos might settle along the Mekong River for access to water. Settlements in Cambodia might be spread out across lowland areas.)*
2. **Monitor Comprehension** What type of severe weather hits mainland countries? *(typhoons)* What environmental danger do island countries face? *(tsunamis)*

Interpret Maps Direct students to the Southeast Asia Physical map or project it from the **Online World Atlas.** Have volunteers trace around the outside of the islands that make up Indonesia and the Philippines. **ASK:** In general, in what ways are the countries different in terms of physical geography? *(Indonesia has more land area and higher elevations; the islands that make up the Philippines are generally smaller than those of Indonesia. Indonesia shares an island with another country.)*
0:15 minutes

DIFFERENTIATE

Connect to NG

Striving Readers Outline and Take Notes Give students the form for an outline such as the one shown here. Explain that the blue subheadings represent the main sections of the lesson. Tell them to use the lines provided to record the main ideas under each subheading. They can record supporting details for each main idea on their papers to the right of the outline.

Outline

I. ____________________
  A. ____________________
  B. ____________________
II. ____________________
  A. ____________________
  B. ____________________

Pre-AP Compare Climates Have students create a chart that compares the climate of a Southeast Asian country with the climate of their own area. Suggest that they use the **Research Links** to find information about each country's climate, such as temperature variations, precipitation levels, and severe weather patterns.

ASK: What generalization can you make about how life would be different in the two areas because of climate? *(Possible response: The tropical climate of Southeast Asia would mean that people would not see the change of seasons that people experience in my part of North America.)*

ACTIVE OPTIONS

Interactive Map Tool

Analyze Settlement Patterns

PURPOSE Understand the relationship between settlement patterns and physical geography

SET-UP

1. Open the **Interactive Map Tool**. Set the "Map Mode" to Topographic and zoom in on Southeast Asia.
2. Under "Human Systems—Populations & Cultures," turn on the Population Density layer. Set the transparency level to about 50 percent.

ACTIVITY

1. Have volunteers come up to the interactive whiteboard and trace the outlines of Laos and Cambodia and describe the settlement patterns of each. *(Settlements in Laos lie mostly along the Mekong River. In Cambodia, settlements are concentrated along the river system through the center of the country.)*
2. Assign pairs to explore settlement patterns in another Southeast Asian country and draw a conclusion about the relationship between these patterns and physical geography.
3. Have pairs point out the physical features and settlement patterns that led them to their conclusion. *(Responses will vary, but students are likely to point out the concentration of settlements around water systems and lowland areas.)* 0:25 minutes

On Your Feet

Talk on Topic Label four areas of the room: Ocean, Mountains, Rivers, Severe Weather. Direct students to the area with the geographic feature they think has the greatest impact on people in the region. Then have students in each area work together to create a visual that illustrates the geographic feature and its impact. Have groups present their visuals. Keep the visuals available to refer to during reading. 0:20 minutes

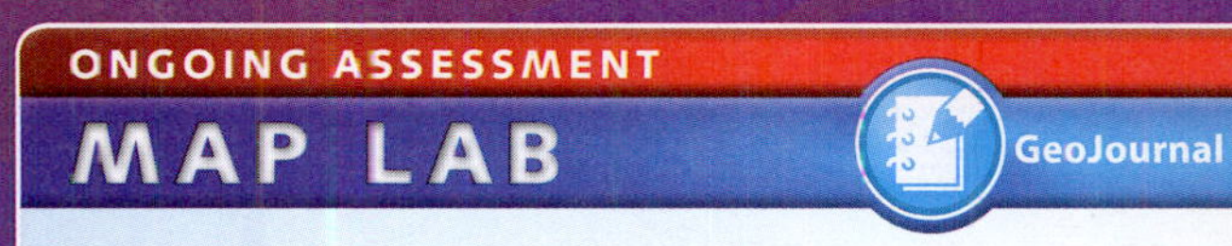

ANSWERS

1. Some countries are located on the mainland of Asia, and others are made up of islands.
2. Difficulty—transportation might be more difficult without a waterway to the ocean. Benefit—it is likely to be damaged by a tsunami.
3. The fertile soil is good for farming, so people can grow food to survive and make a living.

SECTION 1 GEOGRAPHY

1.2 Parallel Rivers

TECHTREK myNGconnect.com For a map and photos of Southeast Asia's rivers
Maps and Graphs
Digital Library

Main Idea River systems in Southeast Asia support life in many ways.

Three parallel rivers run through mainland Southeast Asia: the Mekong (may KONG), the Chao Phraya (chow PRY uh), and the Irrawaddy. They begin in the highlands and flow south through valleys between mountains. As they approach the sea, they divide into a triangular shape made up of smaller streams. These river deltas are composed of silt, or fertile soil the rivers carried from upstream.

The Mekong River

At 2,600 miles, the **Mekong River** 1 is the longest in Southeast Asia. It runs through the middle of the mainland and forms part of the borders of Myanmar, Laos, and Thailand. The mouth of the river, where it empties into the South China Sea, is in Vietnam near Ho Chi Minh City.

The Mekong Delta covers nearly 25,000 square miles, about the size of West Virginia. The densely populated delta is a rich rice-growing region. Some countries in the region are working to harness the river's power to produce hydroelectricity.

The Chao Phraya River

The **Chao Phraya River** 2 is the most important river in Thailand. It is used to irrigate rice fields and serves as a major transportation route through the country. The capital city of Bangkok is located along its banks.

The Irrawaddy River

The **Irrawaddy River** 3 is about half as long as the Mekong. It also supports rice farming and is used as a transportation network. As a result of the soil carried by the river and dumped at its mouth, the delta is growing by about 165 feet a year.

Visual Vocabulary An **ecologist** is a scientist who studies the relationship between organisms and their environments. National Geographic Fellow Zeb Hogan is an aquatic ecologist working here in the waters of the Mekong River.

RIVERS AND POPULATION DENSITY

MAP TIP The land near river deltas is very fertile. Cities grew large on this land because they could feed the population. Except for landlocked Laos, each capital city in the region is located on a river delta.

Population Density
One dot represents 100,000 people

In the rainy season, the Irrawaddy can rise more than 30 feet. Ports must have two areas for docking, one for each season. Farmers have adapted by storing water during the rainy season and releasing it onto their fields in the dry season.

Challenges of the Rivers

The rivers of Southeast Asia are used by many people for **subsistence fishing**, catching just enough fish to live on. Countries in the region must work together to control the threat of overfishing. **Ecologist** Zeb Hogan is part of a program that buys live fish from local fishermen for study. Efforts like this can protect endangered fish while still allowing local people to make a living.

Dams built along the rivers help control water levels, but they also sometimes interfere with transportation and disrupt river environments.

Before You Move On

Summarize What do river systems provide to support life in the region? fertile land in the delta, fish for food and farming, transportation

ONGOING ASSESSMENT

READING LAB GeoJournal

1. **Monitor Comprehension** How are dams a threat to the region's rivers?
2. **Make Inferences** Why is water level a challenge for people who depend on the rivers?
3. **Region** Trace the course of the Mekong River on the map. What countries share this river?

PLAN

OBJECTIVE Understand the importance of Southeast Asia's major rivers for sustaining life in the region.

CRITICAL THINKING SKILLS FOR SECTION 1.2

- Main Idea
- Summarize
- Monitor Comprehension
- Make Inferences
- Compare and Contrast
- Draw Conclusions

PRINT RESOURCES

Teacher's Edition Resource Bank

- Reading and Note-Taking: Organize Information
- Vocabulary Practice: Definition Chart
- **GeoActivity** Research an Environmental Issue

BACKGROUND FOR THE TEACHER

The Mekong River is the longest river in the region and has the largest drainage basin. The Mekong drains more than 313,000 square miles of land, nearly twice the area of the Irrawaddy's drainage basin, which is less than 160,000 square miles. The Chao Phraya, though important to Thailand, covers a much smaller area, about 62,000 square miles.

ESSENTIAL QUESTION

What are the geographic conditions that divide Southeast Asia into many different parts?

Several important rivers course through mainland Southeast Asia. Section 1.2 explains the importance of these rivers—and some challenges facing people who depend on them.

INTRODUCE & ENGAGE

Digital Library

Identify Uses and Threats Show photos of the rivers in Southeast Asia from the **NG Photo Gallery.** Distribute a T-Chart like the one shown, or have students copy the chart onto their own papers. Then give them some time to identify examples in each category and record them on their charts. At the conclusion of the lesson, return to this discussion to verify and add to chart entries. 0:15 minutes

Uses	Threats

TEACH

Guided Discussion

1. **Compare and Contrast** How are the Mekong, Irrawaddy, and Chao Phraya rivers similar? *(They rise in northern highlands and flow south through valleys; they fan out into deltas at the mouth; they support farming, provide fisheries, and are used for transportation; the volume of water they carry varies between the dry and rainy seasons.)* How are they different? *(They differ in length and in the size of the deltas.)*
2. **Draw Conclusions** How would changing water levels affect people who farm lowland areas along the banks of these rivers? Why? *(Changing water levels could mean that some land would be flooded in the rainy season and that more land would be uncovered by water in the dry season. Farmers would have to plant their crops to be harvested before the rainy season.)*

MORE INFORMATION

Fresh water is being used up in large quantities all over the world. One result of this overuse is a decline in aquatic species, and the largest of the freshwater fish seem to be in the most danger. The first expedition of the Megafishes Project, a global conservation effort led by aquatic ecologist Dr. Zeb Hogan, took place in the Mekong River in 2006. Hogan and his team have worked with governments of some Southeast Asian countries to make it illegal to capture the Mekong giant catfish. Conservation areas have been set up in Cambodia to protect large freshwater species such as the giant stingray. Still, many aquatic species are threatened by overfishing and the construction of dam projects in waterways where they live.

DIFFERENTIATE

Connect to NG

English Language Learners **Find Main Idea and Details** Have students work in pairs and assign each pair a paragraph from the lesson. Have them use a chart like the one shown to identify key details in their paragraph. Then tell them to work together to write the main idea based on those details. Have pairs practice their pronunciation by reading the details and main idea sentences aloud.

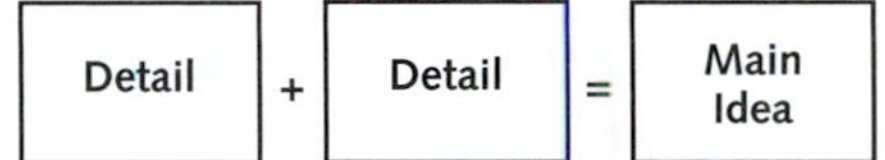

Gifted & Talented **Create Sketch Maps** Have students create a sketch map that shows the countries of mainland Southeast Asia and the three rivers profiled in the lesson. Have them use the **Research Links** to find out the location of current and future dam projects and draw these on the map. **ASK:** Which of these dams might lead to an international dispute? Why? *(Possible response: A dam built near the source of the Mekong River might lead to an international dispute because that river, unlike the other two, flows through more than one country.)*

ACTIVE OPTIONS

Connect to NG

Form and Support Opinions Emphasize that the building of dams in Southeast Asia, specifically on the Mekong River, is very controversial. While the electricity from the dams can boost the region's economy, specifically in Laos, environmentalists believe the dam will destroy marine life—and quite possibly the livelihood of local fishermen.

Have students use the **Research Links** to connect to a National Geographic news story about the Xayabur Dam project. After reading the article with students, have them form groups of four or five. Assign one of the following perspectives to each group, and have them explain the issues from that perspective:

- the Laotian government;
- a citizen of Laos living near where the dam will be built;
- a member of the MRC;
- a member of the World Wildlife Federation;
- a fisherman in Cambodia.

Have each group present their assigned perspective. Then have students vote on whether the building of the dam should go forward. 0:25 minutes

On Your Feet

Turn and Talk on Topic Divide the class into six groups. Give three groups this topic sentence: *The rivers of Southeast Asia provide benefits.* Give three groups this topic sentence: *The rivers of Southeast Asia pose challenges.* Tell the groups to build a paragraph on their assigned topic by having each student add one sentence. Then have groups present their paragraphs to the class with each student reading his or her sentence. 0:15 minutes

ONGOING ASSESSMENT

READING LAB

GeoJournal

ANSWERS

1. The dams interfere with transportation and may also disrupt marine ecosystems and the lives of people who depend on the river.
2. Possible response: Water levels vary dramatically between the rainy and dry seasons, which challenges farmers who need water for their crops and port cities, which need two sets of port facilities.
3. Myanmar, Thailand, Laos, Cambodia, and Vietnam

1.3 The Malay Peninsula

TECHTREK
myNGconnect.com For maps and photos of the Malay Peninsula
Maps and Graphs
Digital Library

Main Idea The mountains of the Malay Peninsula are rich in mineral resources and valuable rain forest land.

The **Malay Peninsula** is long and narrow, only about 200 miles across at its widest point. It includes parts of Malaysia, Thailand, and Myanmar. Mountain ranges rich in mineral resources run the length of the peninsula, and lush rain forest provides a habitat for thousands of plant and animal species.

Mountains and Mining

The Bilaukataung Range mountains of Thailand and the Main Range mountains of western Malaysia have traditionally been mined for tin, a metal often used in food containers. **Bauxite**, the raw material used to make aluminum, is mined in the southern part of the Main Range. Since the 1970s, the number of easily accessed tin and bauxite deposits has been shrinking, causing a steady decline in mining.

Rain Forest, Rubber, and Palm Oil

The peninsula also includes an extensive rain forest, which covers about 40 percent of the land area. The rain forest provides ideal habitat to hundreds of different trees and other plants. This variety of species in an ecosytem is called **biodiversity**. Animals range from large creatures such as elephants, rhinos, and tigers to the very small deer mouse.

Some of the trees native to the rain forest have significant value, and as a result large areas of rain forest have been cleared to plant only those species. At one time, teak wood from Thailand was a large part of the country's economy. However, after a landslide in 1989, which was blamed on excessive deforestation, the government imposed a ban on harvesting teak. In Malaysia, the rain forest is cut down to make room for large farms of rubber and palm oil trees. Palm oil is used with machinery, to make soap, and for cooking.

Critical Viewing This land in Malaysia is being prepared for a palm oil farm. What can you infer was present before the land was cleared? trees of the rain forest

Village on a hillside in Cameron Highlands, Malaysia

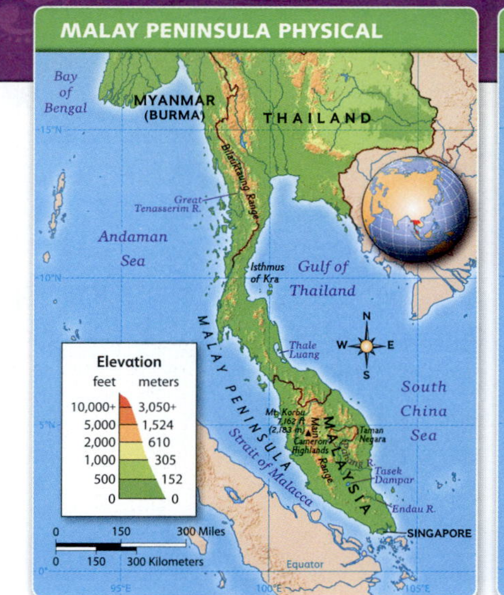

Critics claim that after depleting its mineral resources, Malaysia is now destroying its forests. To clear the land, farmers first cut the trees and then burn anything leftover. Fertilizers are then applied, permanently changing the soil and making it impossible for the rain forest to regrow quickly.

The environmental damage has an economic impact. Beginning in the early 2000s, rain forest tourism became an important part of the peninsula's economy. As a result governments are working to find a balance between land development and rain forest conservation.

Before You Move On

Summarize How do environmental concerns conflict with the production of palm oil?

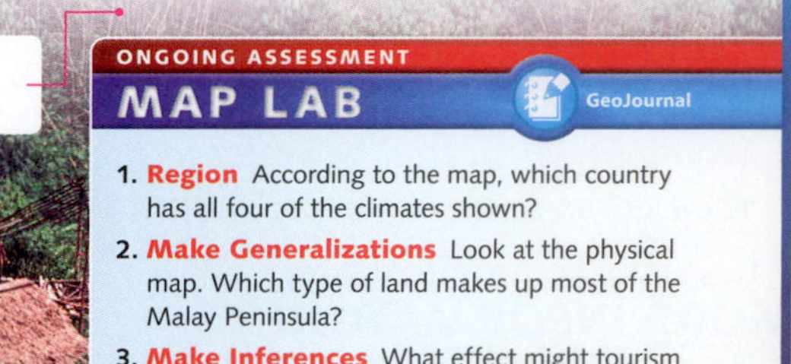

The production of palm oil requires the clearing of rain forest land, which destroys rainforest habitats.

ONGOING ASSESSMENT

MAP LAB

GeoJournal

1. **Region** According to the map, which country has all four of the climates shown?
2. **Make Generalizations** Look at the physical map. Which type of land makes up most of the Malay Peninsula?
3. **Make Inferences** What effect might tourism have on the clearing of rain forest land?

PLAN

OBJECTIVE Examine the main features that characterize the Malay Peninsula.

CRITICAL THINKING SKILLS FOR SECTION 1.3

- Main Idea
- Summarize
- Make Generalizations
- Make Inferences
- Synthesize
- Form and Support Opinions
- Analyze Cause and Effect
- Interpret Maps

PRINT RESOURCES

Teacher's Edition Resource Bank

- Reading and Note-Taking: Find Main Idea and Details
- Vocabulary Practice: Word Map
- **GeoActivity** Graph Global Deforestation Rates

BACKGROUND FOR THE TEACHER

Taman Negara National Park was set aside as a park by Malaysia in 1938. It is Malaysia's largest park and covers land in three of the country's states. The park is reached by taking a three-hour bus ride from Kuala Lumpur, the nation's capital, to a town farther north followed by a three-hour boat trip upriver to a village on the edge of the park.

ESSENTIAL QUESTION

What are the geographic conditions that divide Southeast Asia into many different parts?

Malaysia, like many tropical countries, has a rich and diverse ecosystem in extensive tropical rain forest. Section 1.3 explains the difficulties involved in using and managing that resource.

INTRODUCE & ENGAGE

Use Word Parts Post the word part *bio-*. Have students brainstorm words they know or have heard that include this word part. Some examples might include *biology*, *biography*, and *autobiography*. Inform students that this word part means "life." Ask a volunteer to remind the class of the meaning of *diversity* (variety). Post the word *biodiversity*. **ASK:** What do you think the word *biodiversity* means? *(a variety of life, or types of animals and plants)* Explain that the rain forests of Southeast Asia are known for their biodiversity. 0:10 minutes

bio (life) + diversity (variety) = a variety of life

TEACH

Maps and Graphs

Guided Discussion

1. **Synthesize** How would you describe the shape of mainland Malaysia? *(long and narrow)* How would you describe the geography of that area? *(coastal lowlands rising to hills rising to high mountains in the center)* How is the shape of the Malay Peninsula related to its physical geography? *(The long, narrow shape consists of miles of coastal lowlands on both sides. Higher elevations occur inland or along the center of the narrow peninsula.)*
2. **Form and Support Opinions** Do you think the rain forest in Malaysia should be cleared to make way for more farming? Why or why not? *(Responses will vary. Students may believe farming will create more jobs and improve the economy or that the rain forest should be protected for its unique biodiversity.)*
3. **Analyze Cause and Effect** Why is rain forest land cleared in Malaysia, and what are the effects? *(Land is cleared for its valuable wood and to make way for commercial farming. The effects include loss of habitat, landslides, and damage to the soil.)*

Interpret Maps Help students interpret the Climate map. Show the Climate map from the **Online World Atlas.** Direct their attention to the map legend. Ensure that students understand the differences among the three types of humid equatorial climates. **ASK:** Why is Malaysia's climate suitable for promoting the growth of a rain forest? *(Since there is no dry season, precipitation will fall year round.)* 0:15 minutes

DIFFERENTIATE

Digital Library Connect to NG

Inclusion Analyze Visuals Pair students of mixed ability. Show photos of the Malay Peninsula from the **NG Photo Gallery**. Pause long enough at each photo for students to work together to match the photo to details in the lesson. Have each pair use these details to write a caption for each photo. Then show the photos again, and have pairs share their captions.

Mount Tahan, Malay Peninsula

Gifted & Talented Describe the Rain Forest Have students use the **Research Links** to learn more about Malaysia's rain forest. Instruct them to choose a way to describe that ecosystem. They might create a multimedia display showing the plants and animals that live in it, draw pictures of those species, make a model, or write a song or poem describing the environment. **ASK:** What is the value of keeping a region like this rain forest in its natural state? How important should that be to Malaysia as a goal? Why? *(Responses will vary but may include the value of biodiversity or the medicinal resources being found in the rain forest)*

ACTIVE OPTIONS

Interactive Whiteboard

GeoActivity

Graph Global Deforestation Rates Have students complete the graph in small groups to understand the conflict between protecting resources and economic development. Monitor groups as they plot points on their graphs. Allow groups to compare their results. Encourage them to reconcile differences in the graphs. As a class, discuss the graphs and then analyze the steps that Malaysia might take to address the issue of deforestation. 0:20 minutes

Connect to NG

Extend Your Understanding Have students use the **Research Links** to connect to a National Geographic news story about efforts being made to protect the rain forest on the island of Borneo, part of which is the island portion of Malaysia. **ASK:** What is the government doing to protect rain forest land? *(issuing permits, enforcing boundaries and regulations, offering financial incentives)* Students may wish to further explore the efforts of other agencies to protect the Malaysian rain forest. 0:20 minutes

On Your Feet

Debate Solutions Divide the class into two groups: those who believe that the rain forest land should be used for further economic development and those who believe in protecting the rain forest. Have the students in each group agree on the strongest arguments in favor of their position. Each group should also generate some suggestions to address the concerns of the other group. Then give the two groups a chance to debate the topic by taking turns offering their suggestions. Encourage the class to reach consensus on one or two suggestions. 0:25 minutes

ONGOING ASSESSMENT

MAP LAB

GeoJournal

ANSWERS

1. Thailand
2. Possible response: Central highlands running north to south with coastal plains to the east and west
3. Possible response: If the government decides that rain forest tourism brings in enough money, it might slow the rate at which trees are cut down.

SECTION 1 GEOGRAPHY

1.4 Island Nations

TECHTREK myNGconnect.com For maps and photos of the island nations of Southeast Asia

Maps and Graphs | Digital Library

Main Idea Geographic conditions on the islands affect settlement in Southeast Asia.

Five countries of Southeast Asia are islands or groups of islands: Indonesia, Singapore, Brunei, East Timor, and the Philippines. A part of Malyasia is on the island of Borneo. Mountains and water barriers among the islands have given rise to isolated cultures with distinct features.

Volcanic Activity

Located where the Eurasian, Indian, Philippine, and Australian plates come together, the islands of Southeast Asia are in a **dynamic**, or continuously changing, geographic zone. These plates are constantly—yet slowly—moving into and over each other. Most islands in the region were formed by the crashing together of these plates. The collisions gradually formed small land masses with high volcanic mountains that slope downward toward coastal plains.

Although many of the volcanoes above ground are no longer active, some that are **dormant**, or inactive for long periods of time, can suddenly erupt. For example, Mount Sinabung on Sumatra in Indonesia had been quiet for 400 years before erupting in 2010. Although many farmers stayed with their farms, tens of thousands of people fled. Volcanic eruptions can destroy villages, but the ash also creates the fertile, nutrient-rich soil that allows for successful farming. The humid climate of the islands **enhances**, or improves, the quality of agriculture. As a result, many crops can be cultivated year-round.

Indonesia

Indonesia is the giant of the region in both land area and population. It is nearly three times the size of Myanmar, the next largest. In population, Indonesia is more than three times as large as the Philippines, and it has more volcanoes than any other country in the world.

Critical Viewing This village sits less than two miles from Mount Batur, an active volcano in Indonesia. What might be some advantages and disadvantages of living in this location?

advantages: fertile soil, access to water; disadvantages: danger of volcanic eruption

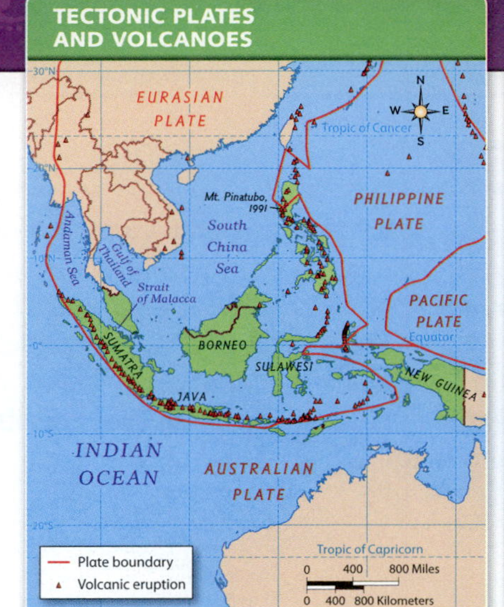

Indonesia is made up of thousands of islands. The five largest islands in size are Sumatra, Java, Borneo, Sulawesi, and New Guinea. Though Java is the smallest of the five, it is also the most populous, with more than half of Indonesia's 240 million people. Four of Indonesia's five largest cities are on Java, including the capital, Jakarta. With the exception of the largest cities, most urban areas are more like large towns, each with its own local culture.

The Philippines

Settlement patterns in the Philippines are similar to those in Indonesia. The greatest concentration of people is on lowland plains, which provide soil made fertile by volcanic eruptions. Like Indonesia, the Philippines has an extremely large capital city. This city, Manila, has more than ten million people.

The Philippines also has many small rural settlements that subsist on fishing or rice farming. Houses near the ocean are built on columns made of timber to allow for changing tides and boat traffic.

Before You Move On

Summarize In what ways has geography of the island nations affected life in Southeast Asia? Volcanic activity has resulted in the formation of many islands.

ONGOING ASSESSMENT

MAP LAB GeoJournal

1. **Place** Based on the map of tectonic plates, which large island seems to be unaffected by active volcanoes or earthquakes? Why might this be?
2. **Compare** Based on the Population Density map, how does the population density of Borneo compare to that of Java? What might be the cause of this difference in population?

PLAN

OBJECTIVE Draw conclusions about how geographic conditions affect life in the island nations.

CRITICAL THINKING SKILLS FOR SECTION 1.4

- Main Idea
- Summarize
- Compare and Contrast
- Make Inferences
- Interpret Maps

PRINT RESOURCES

Teacher's Edition Resource Bank

- Reading and Note-Taking: Outline and Take Notes
- Vocabulary Practice: Comparison Chart
- **GeoActivity** Analyze the Effects of Krakatoa

TECHTREK myNGconnect.com

Fast Forward!

Core Content Presentations

Teach *Island Nations*

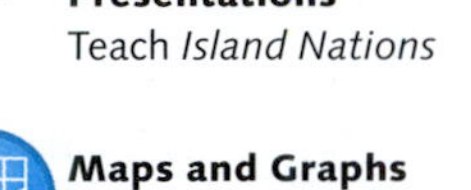

Maps and Graphs

Online World Atlas: Tectonic Plates and Volcanoes; Population Density

Interactive Whiteboard

GeoActivity Analyze the Effects of Krakatoa

Also Check Out

- NG Photo Gallery in **Digital Library**
- Graphic Organizers in **Teacher Resources**
- GeoJournal in **Student eEdition**

BACKGROUND FOR THE TEACHER

Indonesia includes around 17,500 islands. However, nearly 75 percent of the country's area is made up of Sumatra, Kalimantan (the Indonesian portion of Borneo), and Papua (the Indonesian portion of New Guinea). The Philippines has about 7,100 islands, but almost 60 percent of them have no name. The Philippines is slightly larger than Indonesia from north to south, but is only about 700 miles wide at the widest point compared to about 3,200 miles for Indonesia.

ESSENTIAL QUESTION

What are the geographic conditions that divide Southeast Asia into many different parts?

The island nations of Southeast Asia comprise many different land areas. Section 1.4 discusses the characteristics of two of these island nations.

INTRODUCE & ENGAGE

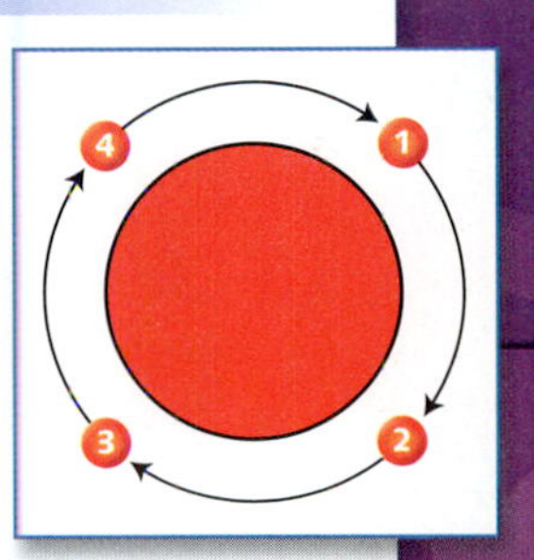

Teamwork: Identifying Issues Have students sit in groups of four as shown in the diagram. Give each group this sentence starter: *The challenges that confront a country that is made up of islands . . .* Give each group time to add some endings to the sentence. Suggest that one group member act as the recorder. Then invite recorders to take turns stating one of their group's conclusions. At the conclusion of the lesson, return to this activity to verify and add information. 0:15 **minutes**

TEACH

Maps and Graphs

Guided Discussion

1. **Compare and Contrast** What are the similarities and differences in the settlement patterns of these two island countries? *(Similarities—most people live in lowland areas; large capital city. Differences—Indonesia has large towns; Philippines has many small rural settlements.)* What might account for the differences? *(Possible response: Indonesia has more large islands; most islands of the Philippines are small.)*
2. **Make Inferences** Why do you think most urban areas in Indonesia have their own distinct culture? *(These urban areas are divided physically and not in contact with one another, which means there is no interaction or blending of culture.)*

Interpret Maps Direct students' attention to the Tectonic Plates and Volcanoes map in the **Online World Atlas**. **ASK:** Where are most of the volcanoes located? *(along the plate boundaries)* What is the reason for the location of the volcanoes? *(As the plates move and collide, volcanic mountains are formed.)* Will the soil found on an island with fewer volcanoes be similar to the soil found on volcanic islands? *(Possible response: The soil on nonvolcanic islands will probably not be as fertile because of the lack of volcanic activity to enrich it.)* 0:15 **minutes**

DIFFERENTIATE

English Language Learners Word Squares Have students create separate Word Squares for the vocabulary words in the text and for at least three other words that are confusing or interesting. Encourage students to write the meaning of the words and to include examples and related words. For example, a Word Square for *dormant* might include words and phrases such as those shown below:

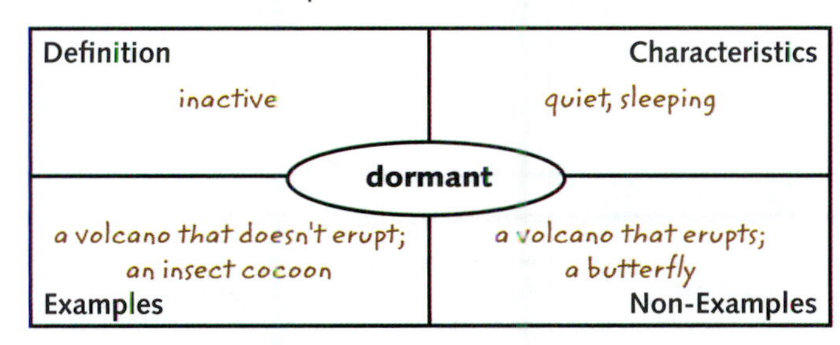

Striving Readers Compare Countries Have students use a Venn Diagram to take notes on Indonesia and the Philippines. Explain that facts unique to either country should be entered in the outer circles with common characteristics placed in the overlapping area. When students have finished, have them make a generalization about the two nations. *(Possible response: Indonesia and the Philippines differ in size, shape, and number of islands, but their climates, location in tectonic danger zones, and patterns of population density are similar.)* Have students complete this activity in preparation for the On Your Feet activity at right.

ACTIVE OPTIONS

Interactive Whiteboard

GeoActivity

Analyze the Effects of Krakatoa Have students complete the Cause-and-Effect map in the activity. After students work individually, have them meet in small groups to review their completed maps. Give students the opportunity to revise their maps to amplify or clarify their responses based on these conversations. As a class, discuss why people choose to live in areas like Indonesia and the Philippines despite the physical dangers. 0:15 **minutes**

On Your Feet

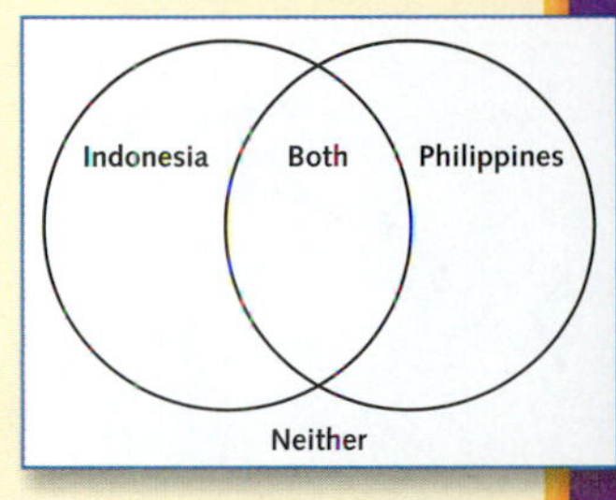

Create a Living Venn Diagram Have each student write one fact on a slip of paper. The fact should be true about Indonesia, the Philippines, both countries, or neither country. Place the slips of paper in a hat. Designate one side of the room as "Indonesia," the other side as "the Philippines," and the middle as "Both." Select a space off to the side of the room for "Neither." One by one have students select a slip of paper, read the fact aloud, and then move to the appropriate area of the room. After all the slips have been read, have each group check each others' facts for correctness and relocate if necessary. Then have groups summarize their facts for the class. Be sure to discuss why the "Neither" facts don't apply to either country. 0:25 **minutes**

ONGOING ASSESSMENT

MAP LAB

GeoJournal

ANSWERS

1. Borneo; it is not located on a plate boundary
2. Java's population density is far higher. Possible response: Java's soil is far more fertile because of the presence of so many volcanoes; Borneo's soil is less volcanic and less fertile.

TECHTREK
myNGconnect.com For a map of New Guinea, photos, and an Explorer Video Clip
Maps and Graphs
Digital Library

Discovering New Species with Kristofer Helgen

Main Idea The unexplored Foja Mountains of Indonesia may be home to unidentified plant and animal species.

Undiscovered Species

Today knowledge of animal and plant life is extensive and well documented. However, National Geographic Emerging Explorer Kristofer Helgen is a **zoologist**, a scientist who studies animals, and he knows there are hundreds of species that have yet to be discovered.

The Foja Mountains

The **Foja Mountains** of Indonesia have wildlife not found in any other region. The area is also largely unexplored. Helgen described the Foja (FOY ya) Mountains as "one of the few places in the world with no villages, no roads, no human population at all." Human presence is not part of life in this rain forest. "The animals there just don't know people," he said. "That's a very, very rare thing in this day and age."

In 2005, Helgen took part in a research trip to this remote area. The team included Helgen, who studies mammals, and other experts on plants, butterflies, reptiles, and birds.

The researchers found 20 new frog species, 5 new kinds of butterfly, and several new plant species. Helgen himself found new rat and mouse species. He also discovered a unique type of **wallaby**, a small relative of the kangaroo, which he hopes to name as a new species.

myNGconnect.com For more on Kristofer Helgen in the field today

New species of blossom bat

Canopy of the Foja Mountain rain forest

"Pinocchio" frog, a new species of tree frog

New species of wallaby may be smallest ever

Scientific Opportunity

The research conducted in the Foja Mountains provides a chance for scientists to learn more about the diversity of animal and plant life. Because they have not been explored, the mountains also offer new scientists an opportunity to become trained in the discovery of new species. Students can join experienced scientists in future trips to the region. They will learn about working in the field even as they help make new discoveries.

Before You Move On

Make Inferences Why might there be many undiscovered species of plants and animals in the Foja Mountains? **The Foja Mountains have no human presence, so there might be animals no one has ever discovered.**

ONGOING ASSESSMENT

VIEWING LAB

GeoJournal

1. **Analyze Visuals** Go to the **Digital Library** and view the Explorer Video Clip about Kristofer Helgen. How would you describe the land?
2. **Place** What in the video explains why scientists are excited about this area?
3. **Make Inferences** How do you think animals that have never encountered humans would behave the first time they see a person? Why?
4. **Create Charts** Make a chart listing the kinds of new species that might be found in a place like the Foja Mountains. Remember that the region is home to a rain forest.

PLANTS	ANIMALS

PLAN

OBJECTIVE Analyze the importance of unexplored areas of Southeast Asia.

CRITICAL THINKING SKILLS FOR SECTION 1.5

- Main Idea
- Make Inferences
- Analyze Visuals
- Create Charts
- Make Predictions
- Create Graphs

PRINT RESOURCES

Teacher's Edition Resource Bank

- Reading and Note-Taking: Summarize Information
- Vocabulary Practice: Blog Entry
- **GeoActivity** Investigate New Species

TECHTREK myNGconnect.com

Fast Forward!
Core Content Presentations
Teach *Discovering New Species*

Digital Library
Explorer Video Clip: *Kristofer Helgen*

Connect to NG
Research Links

Interactive Whiteboard
GeoActivity Investigate New Species

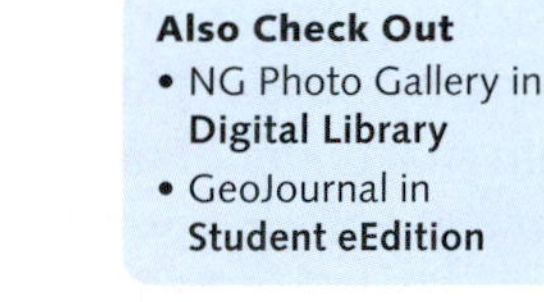

Also Check Out
- NG Photo Gallery in **Digital Library**
- GeoJournal in **Student eEdition**

BACKGROUND FOR THE TEACHER

Field work has taken Helgen not only to the Foja Mountains of New Guinea but also to the Andes Mountains of South America, where he identified a new species of olingo, a cousin of the raccoon. Other field work put him in the Caribbean, where he concluded that a particular kind of raccoon was not a native species and was threatening native birds and sea turtles.

ESSENTIAL QUESTION

What are the geographic conditions that divide Southeast Asia into many different parts?

Living things can develop in unique ways on isolated islands. Section 1.5 explains the causes —and results—of the isolation of a mountain chain in New Guinea.

INTRODUCE & ENGAGE

Digital Library

Explorer Video Clip: *Kristofer Helgen* Show the **Explorer Video Clip** to explore the interesting work that Helgen has done in New Guinea. Ask students to describe his work.

Teamwork: Plan an Expedition Have students form groups of five or six. Have them imagine they are a team of people who are planning an expedition to a remote area to search for new plant and animal species. As a team, they must have the right combination of skills and knowledge to make the expedition a success. Have the group discuss and assign a role to each member of the expedition. Then have groups introduce themselves and explain the ways in which each team member will contribute to the success of the expedition. At the conclusion of the lesson, have teams compare themselves to Helgen's team and discuss similarities and differences. 0:10 minutes

TEACH

Guided Discussion

1. **Make Inferences** Why would it be important for a research team looking for new species include experts on different species? *(Experts would be able to identify the small characteristics that indicate a different species in their field.)*
2. **Make Predictions** Suppose you were a scientist preparing a new expedition to the Foja Mountains. What new kinds of experts would you take? Why? *(Possible response: experts on trees and flowering plants to have more specific knowledge of plant life; experts on insects and spiders because the mountains probably have many of those species; experts on fish to study fish in the streams and rivers in the mountains)*

Create Graphs Tell students that the researchers had three different camps, each at a different elevation (the Lower Camp at 4,003 ft, the Bog Camp at 5,577 ft, and the Mountain Camp at 6,480 ft). Have pairs of students plot these elevations on a graph. Then have them annotate each point on the graph with the characteristics of the plant and animal species that might be found at each elevation. Remind students that temperatures get lower as elevation gets higher. Have pairs share their graphs and their ideas about species. **ASK:** Why would they want to conduct studies at three different elevations? *(Different plants and animals would be found at different elevations.)* 0:20 minutes

DIFFERENTIATE

Connect to NG

Inclusion Summarize Information Use a Fishbowl activity to review the lesson. Place students of mixed ability in each circle. Call on more advanced students to take turns summarizing the lesson content. When the first group of students has concluded its summary, switch positions and have the included students review the lesson content.

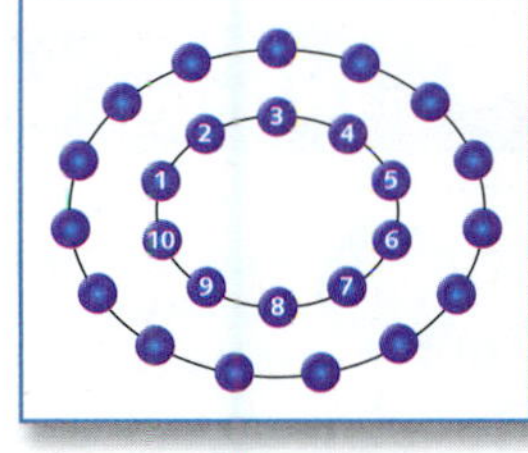

Pre-AP Explore Biodiversity in the Region Have students choose the plant or animal life of one country in the region to study. Using the **Research Links** and reliable Internet sources, have them research and report on one of the following topics:

- What area of the country has a rich diversity of species? What is the area like?
- What plant or animal species unique to the country is particularly notable? What makes it different?

Encourage them to include maps, photographs, and other visuals in their reports.

ACTIVE OPTIONS

Interactive Whiteboard

GeoActivity

Investigate New Species Have students complete the activity in small groups. Pair groups and have them review and discuss their responses. Allow groups time afterwards to expand or revise their responses on the basis of the discussion. As a class, discuss the value of fieldwork to scientific discovery. 0:20 minutes

On Your Feet

Quiz Groups Have students form pairs and write five quiz questions based on the lesson. Each student in the pair should record the questions. Then form new pairs of students and have them take turns quizzing each other. 0:15 minutes

Performance Assessment

Prepare a Museum Exhibit Have teams of three students work together to prepare a museum display that summarizes the geography of Southeast Asia. Have individual students specialize in one aspect of geography, such as landforms, climate, the effects of plate tectonics, resources, or biodiversity. After conducting initial research and formulating ideas for their own contribution, all students in the group should meet to plan the overall look of their display. Go to **myNGconnect.com** for the rubric.

ONGOING ASSESSMENT

VIEWING LAB

GeoJournal

ANSWERS

1. Responses will vary. Students might point to the thick vegetation and mountainous landscape.
2. The Foja Mountains are different because the plants and animals there have not had any human contact. Scientists are excited because they can identify new species of living things never seen before.
3. Responses will vary. Students might suggest that animals would be fearful of human contact.
4. Possible responses: *Plants*—trees, shrubs, flowers, mosses; *Animals*—birds, frogs, lizards, snakes, insects, possibly monkeys, predators to eat other animals

2.1 Ancient Valley Kingdoms

TECHTREK myNGconnect.com For photos of ancient Southeast Asian kingdoms

Digital Library

Main Idea The development of Southeast Asia was influenced by nearby powers because of its important location for trade.

The location of Southeast Asia between the Pacific and Indian Oceans meant its surrounding waterways were on important trade routes. Two powerful civilizations, China to the north and India to the west, heavily infuenced the cultural direction of the region. The impact came through military force and invasion, as well as through trade.

Mainland Empires

Chinese culture first came to Southeast Asia in 111 B.C., when the Chinese invaded and conquered part of what today is Vietnam. India had already established trade before the Chinese arrived. This trade had a strong influence on the area's religious practices.

By the A.D. 700s, Buddhist and Hindu empires were competing for influence and power in other parts of the region. The largest and longest lasting was the **Khmer Empire** of Cambodia. Centered along the Mekong River valley, the Khmer (kuh MAYR) Empire covered much of Southeast Asia and lasted from the A.D. 800s to the 1430s.

At the peak of the empire's success in the 1100s, its ruler, King Suryavarman II, built the massive Hindu temple **Angkor Wat** in the capital city.

Angkor Wat, Cambodia

This religious **complex**, or set of interconnected buildings, was dedicated to the Hindu god Vishnu and served as a tomb for the king. Beautiful **bas-reliefs**, or sculptures that slightly project from a flat background, cover the walls with scenes from Hindu stories. Eventually forces from modern-day Thailand conquered the city and the complex fell into ruin.

In A.D. 939 the people of Vietnam broke from China and established the independent kingdom of Dai Viet. Though influenced by Chinese culture, Vietnam had its own cultural traits. For instance, women in Vietnam had higher social standing than women in China. Eventually Dai Viet grew weak and was reconquered by China in 1407.

Island Empires

Modern Indonesia was also home to powerful empires. The earliest was Srivijaya (sree vi JY ah), which arose on southern Sumatra in the A.D. 600s. This kingdom controlled the Strait of Malacca and was therefore able to control trade from South Asia to China. It was known throughout Asia as a center of trade and also as a center of Buddhist study. The empire declined around A.D. 1100.

Chinese and Indian cultures brought Buddhism and Hinduism to the region, which brought about the building of massive religious temples.

A second Indonesian power was the Sailendra dynasty, which arose in Java and flourished from about A.D. 780 to 850. Sailendran rulers built another famous temple complex, **Borobudur.** Each of the temple's three levels symbolizes a step toward enlightenment, the ultimate spiritual goal of Buddhism.

Another trading kingdom arose on Java around A.D. 1300. It is named for its capital city of Majapahit (mah jah PAH heet) in eastern Java. It gained power through the control of trade. However, by the 1500s, other powers had replaced it.

Before You Move On

Summarize How did Chinese and Indian empires influence life in the region?

ONGOING ASSESSMENT

VIEWING LAB GeoJournal

1. **Interpret Time Lines** Based on the time line, how long was Vietnam able to maintain its independence from China?
2. **Make Generalizations** Look at the photo of Angkor Wat. How would you describe the building skill of the Khmer Empire? Support your answer with details from the photo.
3. **Compare and Contrast** Based on the photos, how are the Borobudur temple and Angkor Wat similar and different?

Borobudur temple, Java, Indonesia

Angkor Wat bas-relief sculpture

PLAN

OBJECTIVE Analyze the role of physical geography in the history and culture of Southeast Asia.

CRITICAL THINKING SKILLS FOR SECTION 2.1

- Main Idea
- Summarize
- Interpret Time Lines
- Make Generalizations
- Compare and Contrast
- Analyze Causes
- Synthesize

PRINT RESOURCES

Teacher's Edition Resource Bank

- Reading and Note-Taking: Categorize Information
- Vocabulary Practice: Travel Brochure
- **GeoActivity** Solve a Puzzle About Ancient Kingdoms

TECHTREK myNGconnect.com

Fast Forward!

Core Content Presentations
Teach *Ancient Valley Kingdoms*

Digital Library
GeoVideo: *Introduce Southeast Asia*

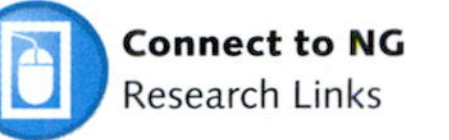

Connect to NG
Research Links

Maps and Graphs
Online World Atlas: Southeast Asia Political

Also Check Out
- NG Photo Gallery in **Digital Library**
- Graphic Organizers in **Teacher Resources**
- GeoJournal in **Student eEdition**

BACKGROUND FOR THE TEACHER

Scientists are still learning new things about Angkor Wat. Scans conducted by the space shuttle *Endeavor* in 1994 and from other remote sensors showed that the complex had hidden buildings and structures that were part of the water management system. Some scientists now suggest that cutting down too many trees and overuse of the land led to excessive flooding and the buildup of silt, which led to the abandonment of the complex.

ESSENTIAL QUESTION

How have physical barriers in Southeast Asia influenced its history?

Mountain ranges in mainland Southeast Asia create isolated river valleys, and islands are remote from each other. Section 2.1 explains how these landforms affected the course of history in the region.

INTRODUCE & ENGAGE

Digital Library

GeoVideo: *Introduce Southeast Asia* Show the beginning of the video that describes the region as a "crossroads for trade" and shows its location on the globe. Explain that a "crossroads" in this context is a meeting place where traders from other regions passed through on their journeys. Have students identify the two oceans on either side of the region. *(Indian, Pacific)* **ASK:** Which other cultures do you think passed through Southeast Asia on their trade routes? *(European, Indian, Chinese)* 0:15 **minutes**

TEACH

Maps and Graphs

Guided Discussion

1. **Analyze Causes** Why would an empire be able to gain power by controlling the waterways? *(because waterways connected larger bodies of water; by controlling them, the empire could control trade)*
2. **Compare and Contrast** What was similar in the spread of Indian and Chinese culture to Southeast Asia? *(Both spread because of the desire to expand trade and culture and/or religion.)* What was different? *(Chinese culture spread by conquest and trade; Indian culture spread by trade.)*

Synthesize Have students work in pairs to understand the repeated conquest of Vietnam by the Chinese. Instruct them to use the time line to identify the two Chinese conquests of Vietnam, and the years of independence between. Show the Southeast Asia Political map from the **Online World Atlas**. Point out China's location north of Vietnam. **ASK:** Why would the Chinese be interested in conquering Vietnam? *(It would give them control of the whole eastern coastline of Southeast Asia.)* 0:15 **minutes**

DIFFERENTIATE

Connect to NG

Inclusion **Summarize** Give students a 5Ws Chart such as the one shown. Ask them to take notes on each kingdom or empire by answering the five questions on the chart. Remind them that answers might be found in more than one place in the text. For instance, the beginning and ending years of a kingdom or empire might be in different paragraphs.

5Ws Chart

What?
Who?
Where?
When?
Why?

Gifted & Talented **Portray Achievements** Have students choose one of the kingdoms or empires discussed in the lesson and use the **Research Links** to find out more about how they lived and what they achieved. Then have them use one of the arts to portray the cultural or historical achievements of that kingdom or empire. They might consider one of these options:

- Drawing or sculpting an image (such as the bas-reliefs of the temples)
- Making a model of a building
- Dancing to music
- Writing a poem or presenting a reading from the literature of the culture

Have them share their presentation with the rest of the class. As a class, discuss what these cultural and historical achievements reveal about the kingdom or empire, and the effects of these achievements. For example, the temple complexes built by ancient kingdoms shows their high level of architectural skill.

ACTIVE OPTIONS

Connect to NG

Categorize Information Have students use the **Research Links** to connect to an online article from *National Geographic* about Angkor Wat. Read the first few paragraphs aloud to the class. Explain that scientists have many theories about what caused the decline of the Khmer kingdom at Angkor Wat. Divide the class into seven or eight small groups and assign each group different sections of the article. Instruct students to take notes as they read, categorizing evidence in the article into the following topics:

- Angkor Wat's water system
- conflict internally and with rival kingdoms
- religious beliefs and practices
- effects of monsoon patterns

After groups finish their assigned section, reassemble groups into the four categories listed above. Have each group explain the evidence presented in the article about their topic. Discuss as a class which theory students believe is the most likely cause of the empire's decline. 0:30 **minutes**

On Your Feet

Analyze Kingdoms and Empires Create four cards labeled "Khmer Empire," "Dai Viet," "Srivijaya Empire," and "Sailendra Empire" and place them on the floor in a rough estimation of their location relative to each other. Have students go to the location of the kingdom or empire that most interests them. Once students are gathered, have them identify the features that make "their" empire unique and powerful, and reasons why it might have failed. Have each group describe their kingdom and offer (possible) reasons for its decline. As a class, evaluate which kingdom's features are most important for a successful civilization. 0:20 **minutes**

ONGOING ASSESSMENT

VIEWING LAB

GeoJournal

ANSWERS

1. 468 years
2. Possible response: The Khmer Empire had skilled builders, which is evident from the beauty and intricate detail of the structures at Angkor Wat and Borobudur.
3. Both have many steps and platforms. Both are built of some sort of stone and show great building skill. Angkor Wat has several taller towers while Borobudur only seems to have one structure protruding from the top.

SECTION 2 HISTORY

2.2 Trade and Colonialism

TECHTREK myNGconnect.com For maps and photos of trade
Maps and Graphs | Digital Library

Main Idea The development of the spice trade in Southeast Asia led to colonization of the region.

As you have learned, trade with India and China brought the influence of these cultures to Southeast Asia. European influence arrived in the 1500s, as merchants hoped to establish a spice trade **monopoly**, or complete control of the market. Spices found in the region, such as cinnamon, nutmeg, and black pepper, could be sold for a high profit in Europe. Traders from Spain and Portugal came first, but the Netherlands' **Dutch East India Company** dominated the region for many years. This success established a strong Dutch influence in Indonesia.

European Control

From the 1600s to the 1800s, Europeans tried to gain an economic hold on Southeast Asia. By 1850, through a combination of alliances, favorable trade agreements, and even military force, the majority of the region was ruled by European powers. (See the time line below.) Only Thailand and parts of the Philippines were independent. Britain, France, Spain, and the Netherlands controlled the rest.

The economic motives that led European powers to Southeast Asia changed the region. Increased production and an ongoing demand for goods strengthened the region's economy. However, trade and wealth, once in the hands of indigenous kingdoms, were now held by distant economic powers. **Colonialism**—one country ruling and developing trade in another country for its own benefit—continued in Southeast Asia well into the 20th century.

Before You Move On

Monitor Comprehension How did development of the spice trade lead to colonization? European countries established colonies in the region to gain control over the profitable spice trade.

SOUTHEAST ASIA UNDER COLONIAL RULE, c. 1895

SPAIN

1521 Ferdinand Magellan lands at the Philippines and claims them for Spain.

1565 Spain makes first settlement on the Philippine islands.

1571 Spain captures site of Manila.

1830s Spain opens Manila to trade.

1892 Filipinos begin movement aimed at independence from Spain.

1898 United States defeats Spain in the Spanish-American War and wins control of Philippines.

NETHERLANDS

1619 Dutch East India Company makes base at Batavia (modern Jakarta).

1641 The Netherlands captures Malacca from Portugal and becomes a major power in spice trade.

1824 The Netherlands and Britain reach agreement on control of Java and Sumatra (to Dutch) and Singapore and Malacca (to British).

1825–1839 The Netherlands fights to put down revolts on Java.

1860 The Netherlands and Portugal sign treaty to divide Timor between them.

GREAT BRITAIN

1781 Britain captures Sumatra from Dutch.

1786–1809 Britain gains control of Malaya trade.

1819 Britain founds Singapore, which becomes a major port city.

1824–1826 Britain controls western Burma.

1886 Britain completes control of Burma in Third Burmese War.

1888 Britain wins southern Burma.

1888 Britain gains control of northern Borneo.

FRANCE

1644 France forms French East India Company.

1789 French East India Company disbanded during French Revolution.

1858 France captures Saigon, Vietnam.

1863 France seizes Cambodia.

1887 France creates Indo-Chinese Union (Cambodia, Vietnam).

1896 France and Britain agree to allow Siam to remain independent to separate their colonies.

ONGOING ASSESSMENT

MAP LAB GeoJournal

1. **Region** Based on the map, which European countries held the most territory in the region around 1895?
2. **Movement** Think about the importance of waterways in influencing trade in this region. Which European country was in the best position to control trade? Support your answer with evidence from the map.
3. **Interpret Time Lines** Based on the dates and events, what do you think was the attitude of the people of Southeast Asia toward European control? Why do you think so?

PLAN

OBJECTIVE Learn about about the effects of colonialism in Southeast Asia.

CRITICAL THINKING SKILLS FOR SECTION 2.2

- Main Idea
- Monitor Comprehension
- Interpret Time Lines
- Draw Conclusions
- Summarize
- Interpret Maps

PRINT RESOURCES

Teacher's Edition Resource Bank

- Reading and Note-Taking: Draw Conclusions
- Vocabulary Practice: Definition Clues
- **GeoActivity** Map the Spice Trade

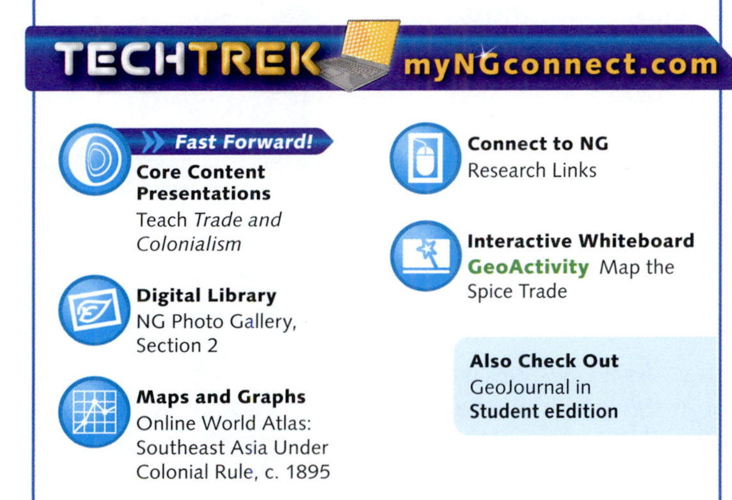

BACKGROUND FOR THE TEACHER

Arabs were the traders of spices from Southeast Asia to Europe through the Middle Ages and the Renaissance. In the 900s, Venice gained control of the trade of spices. However, Europeans still depended on Arab merchants for access to the spices. That changed around 1500, when Portugal gained hold of the trade. It dominated until 1595 when the Dutch became the leading European spice traders. The Dutch were the major power in Southeast Asia until the British supplanted them.

ESSENTIAL QUESTION

How have physical barriers in Southeast Asia influenced its history?

One barrier that blocked European access to Southeast Asia for hundreds of years was the vast travel distances. Section 2.2 explains what happened when that changed.

INTRODUCE & ENGAGE

Digital Library

Experience Spices Have students use an Idea Web to brainstorm foods that contain spices that originated in Southeast Asia. If possible, ask volunteers to bring in samples (or supply your own) of these spices and some foods that contain them. Also show images of the spice trade from the **NG Photo Gallery.**

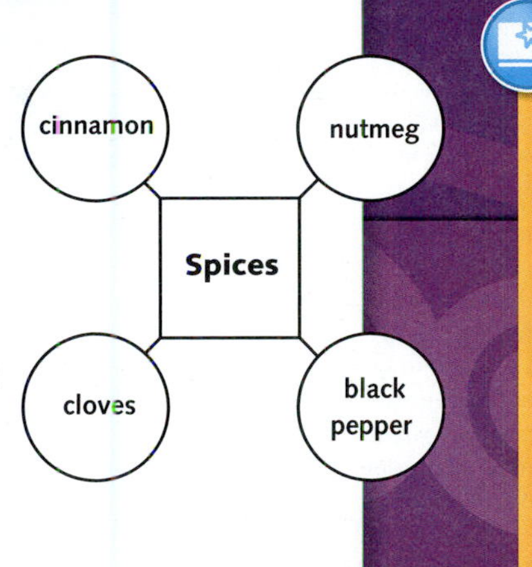

Explain that these spices were highly desirable in Europe and therefore a profitable trade commodity. However, they were only available in Southeast Asia. **ASK:** If these spices never became part of global trade, which foods that you commonly eat today would likely not exist? *(Responses will vary.)* As students sample spices and foods, have them identify which foods they would no longer eat if they were missing spices. 0:15 minutes

TEACH

Maps and Graphs

Guided Discussion

1. **Draw Conclusions** Why did European countries want a monopoly on the spice trade? *(They could earn huge profits from trade in these goods.)*
2. **Summarize** What major changes were brought about in the region due to the development of the spice trade? *(Merchants from Europe established companies to participate in the spice trade; most of the region came under European control; resources in the region were no longer in local control)* Which of these changes do you think were permanent? Explain. *(Responses will vary, but students might point out that colonized countries can gain their independence, but European influence on the region could be permanent.)*

Interpret Maps Show the Southeast Asia Under Colonial Rule map from the **Online World Atlas.** Point out that Siam was the only Southeast Asian country to avoid European colonization. **ASK:** What colonial powers surrounded Siam? *(Britain to the north, west, and south; France to the north, east, and south)* Do you think it would have been easy or difficult for Siam to keep its independence? Why? *(Possible response: It must have been difficult given that Siam was surrounded—and that the European powers were so determined to gain control of the region.)* 0:15 minutes

DIFFERENTIATE

Connect to NG

Striving Readers Make Generalizations Assign pairs to write a summary of one of the four time lines. Suggest that they write their summary in story form. Have pairs share their summaries and use them to review the lesson. You may wish to provide sentence frames:

> *[Country] first made contact in _____ in the year _____. Settlers went on to capture/found _____. By _____, the country had control of _____. Toward the end of the 1800s, _____.*

Pre-AP Research the Spice Trade Have students use the **Research Links** to research the European spice trade. Have them prepare an illustrated report that answers questions such as the following:

- Which areas in Southeast Asia produced which spices?
- Why were the spices so desirable?
- How were they bought and sold?

ACTIVE OPTIONS

Interactive Whiteboard

GeoActivity

Map the Spice Trade Have students complete the map in small groups to study the extent of trade within the region and beyond it. Suggest that groups compare their maps with other groups and discuss any differences. Remind them to refer back to the data when comparing different views of how to map it. As a class, discuss the ways that Southeast Asian and European cultures were impacted by trade. 0:15 minutes

On Your Feet

Debate Positives and Negatives Have each student write a plus sign or a minus sign on an index card. A plus sign indicates a belief that the spice trade had a positive impact on the region. A minus sign indicates a belief that the spice trade had a negative effect on the region. Without showing their cards, have students find a partner and tell each other one fact about the spice trade in Southeast Asia. If the pair believes their cards match, they show their cards. If there is a match, these students stick together and continue to another student. This questioning continues until the class is divided into two opposing groups.

Give each side a few minutes to prepare arguments and then stage a class debate on the issue. Remind students to give facts and reasons to support their opinions. Have speakers from each side alternate. When the debate is done, stage a class vote to determine whether the majority thinks the overall impact was positive or negative. Ask students to discuss how or if the debate changed their opinions about the impact of the spice trade. 0:25 minutes

ONGOING ASSESSMENT

MAP LAB

GeoJournal

ANSWERS

1. Britain, France, and the Netherlands all had large amounts of territory.
2. Responses may vary. Both Britain and the Netherlands were in good positions to control the Strait of Malacca and other important waterways.
3. Possible response: The people of Southeast Asia did not like European control. The people of Java revolted, the Philippines launched an independence movement, and wars were needed by the British to win Burma and the French to win Indochina.

2.3 Indonesia and the Philippines

TECHTREK myNGconnect.com For photos of Indonesia and the Philippines
Digital Library

Main Idea Indonesia and the Philippines are island countries that have faced similar challenges in becoming independent.

As you know, Southeast Asia has a long history of diversity because of its geographic location and unique resources. The island nations of Indonesia and the Philippines have been influenced and even controlled by other cultures throughout history. However, after long struggles for independence, each one has become its own nation.

Indonesia

Indonesia may have been home to the earliest species of humans. **Fossils**, or preserved remains, found on Java suggest that human life existed there as early as 1.7 million years ago. Evidence shows that ancient societies there used hand tools, made implements out of metals, and wove cloth. They also traveled the sea as early as 2500 B.C., possibly to trade with other areas of Asia and beyond.

As its civilization matured, Indonesia became an intersection for trade in the East. Part of the country became known to Europe as the Spice Islands, for the many exotic spices that were a strong attraction for explorers and traders. For example, trade in nutmeg, a spice native to Indonesia, became extremely profitable for the Dutch who had settled there.

Throughout the 1800s, the Dutch expanded their control. Some revolts occurred, but the Dutch were able to maintain power. In 1830, they began a system that required all villages to give part of their crops to the government for export. As the Dutch gained wealth, Indonesians suffered. During the 1900s their resistance efforts became more organized. The Japanese seized control from the Dutch during World War II. As these two countries fought for control, Indonesians continued to resist. At the same time, a strong sense of national identity was developing. The country finally won independence in 1949.

Mace (shown here) is a spice that covers nutmeg in its raw form. Both nutmeg and mace are highly valued for their distinct flavor.

Critical Viewing Emilio Aguinaldo (front, center) and members of the assembly of the First Philippine Republic, 1899. In early photography, subjects were required to sit for long periods to capture an image. How might this explain the expressions on these men's faces?

They have to hold the same expression until the photo is complete. It would be hard to hold a smile for a long period of time.

The Philippines

When the Spanish seized control of **Manila** in 1571, they made it the capital of their new colony. Manila was and still is the economic, political, and cultural center of the Philippines. Trade with China led many Chinese people to settle in Manila. They became the major force in **commerce**, or the business of trading goods and services.

In the 1800s, Spain's economic power began to fade and Manila became open to trade with more countries. This allowed some Filipinos to gain wealth and influence they had not known before.

Emilio Aguinaldo was a key leader in the movement for Filipino independence. He fought alongside the United States when it defeated Spain in the Spanish-American War in 1898, and thought the islands would become independent. However, after the war, the United States kept the islands as its own colony.

In late 1941, during World War II, Japan attacked the United States at Pearl Harbor, then attacked the Philippines. The outnumbered U.S. forces stationed on the islands were forced to leave. Later in the war, the United States took back control and remained in power until 1946, when the Philippines were granted independence. Today, it is a stable democracy.

Before You Move On

Summarize What challenges did Indonesia and the Philippines face in gaining independence?

They both faced the challenge of developing a national identity and creating a stable democracy.

ONGOING ASSESSMENT
READING LAB GeoJournal

1. **Location** Why was Southeast Asia so important to Europeans?
2. **Make Inferences** Why were Indonesians able to develop a strong national identity?
3. **Draw Conclusions** Was trade or conquest more important in shaping these countries? Why?

PLAN

OBJECTIVE Compare and contrast the histories of Indonesia and the Philippines.

CRITICAL THINKING SKILLS FOR SECTION 2.3

- Main Idea
- Summarize
- Make Inferences
- Draw Conclusions
- Identify
- Evaluate
- Synthesize

PRINT RESOURCES

Teacher's Edition Resource Bank

- Reading and Note-Taking: Compare and Contrast
- Vocabulary Practice: Words in Context
- **GeoActivity** Analyze Achievements of Emilio Aguinaldo

Core Content Presentations
Teach *Indonesia and the Philippines*

Interactive Whiteboard
GeoActivity Analyze Achievements of Emilio Aguinaldo

Digital Library
NG Photo Gallery, Section 2

Also Check Out
GeoJournal in **Student eEdition**

BACKGROUND FOR THE TEACHER

The Philippines gained partial self-rule in 1907. In 1935, Congress passed the Tydings-McDuffie Act, which gave more local control to the islands and promised full independence within 10 years. World War II delayed that event. The islands were under Japanese occupation from late 1941 until 1945, though U.S. and Filipino troops returned in 1944 and began winning control of some areas. On July 4, 1946, an autonomous Filipino government took control.

ESSENTIAL QUESTION

How have physical barriers in Southeast Asia influenced its history?

Indonesia and the Philippines are both island nations with large populations. Section 2.3 reviews their colonial history and how they gained independence.

INTRODUCE & ENGAGE

Think, Pair, Share Have students work in pairs to brainstorm reasons that the people of a region might wish to have independence. Members of each pair should think about the issue separately at first and then discuss their different ideas. Tell them to develop a final list of reasons and then invite one member from each pair to share those ideas with the class. **0:15 minutes**

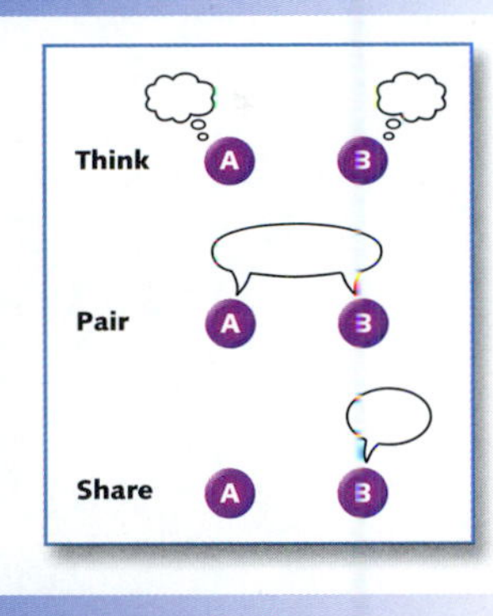

TEACH

Guided Discussion

1. **Identify** What outside countries and cultures influenced Indonesia over the course of history? *(India, China, and Dutch and Muslim cultures)*
2. **Evaluate** How did the contact with those cultures differ? *(Chinese, Indian, and Muslim influence was religious and cultural. All three also engaged in trade with the region. Dutch influence was mainly economic as a result of trade and colonial rule.)*
3. **Make Inferences** Why did Indonesian resistance efforts against the Dutch become more organized during the 1900s? *(The Dutch began a system of taking crops from villages for their own profit. This action may have inspired Indonesians to fight for what was rightfully theirs.)*

Synthesize Assign Indonesia to half the class and the Philippines to the other half. Instruct students to write a short oral presentation that summarizes in their own words the history of their assigned nation. Then have students working on Indonesia partner with students working on the Philippines. Have partners take turns delivering their presentations to each other. **0:15 minutes**

DIFFERENTIATE

English Language Learners Identify Synonyms Have small groups or pairs of students work to identify synonyms for the Key Vocabulary word *commerce*. *(business, industry, trade, traffic)* Remind students to start by looking in the text and then in a dictionary or a thesaurus. Have students write the words in a Word Web, adding words as necessary. Say each word and ask students to repeat after you. Introduce or review the concept of connotations. Tell students that while all of the words are synonyms, they all have slightly different meanings. Have students work with native speakers to add an additional circle to each word with a description of the word's connotation.

commerce

Inclusion Create Time Lines Have students work in pairs to place the events described in the lesson on a time line. Provide pairs with dates for each country: Indonesia: 1800s, 1830, 1900s, 1949; the Philippines: 1800s, 1898, 1941, 1946. Guide students as they place dates and a short description of each date's importance above the time line. For ranges of dates, have them bracket a portion of the time line and write the date and description below it. Post the example shown below. Students may use their time lines to review the lesson.

Time Line

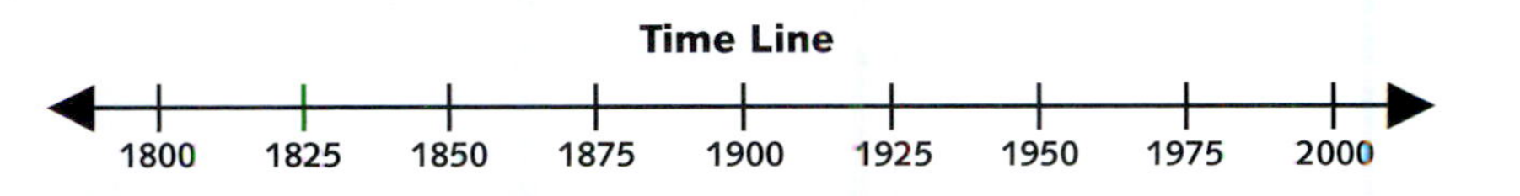

ACTIVE OPTIONS

Interactive Whiteboard GeoActivity

Analyze Achievements of Emilio Aguinaldo Read the passage aloud with students. Have them work in small groups to identify obstacles and achievements. Then have groups compare their charts and make any necessary additions or corrections. **ASK:** Why do you think Aguinaldo's achievements were so important even though he was out of political life when independence was finally achieved? *(His actions throughout his life may have inspired greater numbers of Filipinos to continue resisting colonial rule and become an independent nation.)* Suggest that groups review the lesson before completing Question #2. Have groups share their answers with the class and come to a consensus. **0:20 minutes**

NG Photo Gallery

Make Connections Show the photos of Indonesia and the Philippines. Have students work in pairs to write captions that connect the photo with the history of these two countries. Then show the photos again and have students share their captions as a way of reviewing the lesson. **0:15 minutes**

On Your Feet

Similarities and Differences Divide the class into an even number of groups. Have the groups meet in different areas of the room. Tell half the groups that they are to identify similarities in the histories of Indonesia and the Philippines. The other half is to find differences. Give each group time to prepare a presentation on their topic, with each group member adding a sentence to the presentation. Then give the groups time to present their paragraphs to the class, with each student reading his or her sentence. **0:15 minutes**

ONGOING ASSESSMENT

ANSWERS

1. The region provided spices that were highly desirable and profitable to trade.
2. Indonesia developed a strong national identity by uniting in resistance against the Japanese and Dutch who were fighting for control.
3. Possible response: Trade brought more lasting influences to Indonesia than conquest. Colonial control by the Spanish and Americans had more influence on the Philippines.

SECTION 2 DOCUMENT-BASED QUESTION

2.4 The Vietnam War

TECHTREK myNGconnect.com For photos of the war in Vietnam and Guided Writing

Digital Library · Student Resources

THE VIETNAMESE PEOPLE . . . ARE DETERMINED TO CONTINUE THEIR RESISTANCE UNTIL THEY HAVE WON REAL INDEPENDENCE.

— HO CHI MINH

In 1954 Vietnam was divided into two parts, North and South. President **Ho Chi Minh** of North Vietnam established a Communist government. War erupted in 1959 when he sent aid to overthrow the government in South Vietnam and create one country. The United States supported South Vietnam—it feared that defeat by the North would spread communism to other countries. U.S. forces fought there from 1964 to 1973. In 1975, South Vietnam surrendered. The reunited country became the Socialist Republic of Vietnam.

DOCUMENT 1

President Lyndon Johnson on U.S. Policy

In 1965, Johnson explained reasons for the involvement of the United States in the fighting in Vietnam:

We are . . . there to strengthen world order. Around the globe . . . are people whose well-being rests, in part, on the belief that they can count on us if they are attacked. To leave Viet-Nam . . . would shake the confidence of all these people in the value of an American commitment. . . . Let no one think for a moment that retreat from Viet-Nam would bring an end to conflict. The battle would be renewed in one country and then another. The central lesson of our time is that the appetite of aggression is never satisfied.

CONSTRUCTED RESPONSE

1. How are Johnson's two reasons for fighting the Vietnam War related?

DOCUMENT 2

Letter from Ho Chi Minh

In 1967, Johnson sent Ho Chi Minh a letter suggesting that the two sides begin peace talks. Here is Ho Chi Minh's response:

The Vietnamese people have never done any harm to the United States. But . . . the United States Government has constantly intervened in Viet-Nam, it has launched [started] and intensified the war of aggression in South Viet-Nam for the purpose of prolonging the division of Viet-Nam and of transforming [remaking] South Viet-Nam into an American neo-colony and an American military base. . . .

The Vietnamese people . . . are determined to continue their resistance [opposition] until they have won real independence.

CONSTRUCTED RESPONSE

2. How does Ho Chi Minh respond to the suggestion of peace talks?

DOCUMENT 3

Photo of Warfare

The physical geography and climate of Vietnam presented challenges to soldiers. These challenges might be especially great for soldiers who were not native to the region and were not accustomed to heavy rains and dense plant life of the jungle.

CONSTRUCTED RESPONSE

3. What does the photo suggest about the conditions the soldiers faced in the war in Vietnam?

ONGOING ASSESSMENT

WRITING LAB

GeoJournal

DBQ Practice Think about how North Vietnam compares in size and power to the United States. What factors might have allowed North Vietnam a chance to win the war?

Step 1. Think about Johnson's determination to fight in Vietnam, Ho Chi Minh's statements, and what the photo shows about the war in Vietnam.

Step 2. On your own paper, jot down notes about the main idea expressed in each document.

Document 1: Excerpt: Johnson's Speech
Main Idea(s) ____________

Document 2: Excerpt: Ho Chi Minh's Letter
Main Idea(s) ____________

Document 3: Photo of Warfare
Main Idea(s) ____________

Step 3. Construct a topic sentence that answers this question: What factors might have affected North Vietnam's ability to fight and win a war against South Vietnam and the United States?

Step 4. Write a detailed paragraph that answers the question above, using evidence from the documents. Go to **Student Resources** for Guided Writing support.

PLAN

OBJECTIVE Use primary sources to explore the political reasons for the Vietnam War.

CRITICAL THINKING SKILLS FOR SECTION 2.4

- Summarize
- Analyze Primary Sources
- Identify
- Make Inferences

PRINT RESOURCES

Teacher's Edition Resource Bank

- Reading and Note-Taking: Analyze Primary Sources
- Vocabulary Practice: Word Squares
- **GeoActivity** Compare and Contrast Two Wars in Asia

Core Content Presentations
Teach *The Vietnam War*

Interactive Whiteboard
GeoActivity Compare and Contrast Two Wars in Asia

Connect to NG
Research Links

Also Check Out
- NG Photo Gallery in **Digital Library**
- Graphic Organizers in **Teacher Resources**
- GeoJournal in **Student eEdition**

BACKGROUND FOR THE TEACHER

Ho Chi Minh adopted socialism in the 1910s. He became the leader of a group of Vietnamese patriots and, after World War I ended, sent a petition to the Versailles peace conference demanding more rights for the Vietnamese under French colonial rule. Soon after, he adopted communism with independence for Vietnam as his goal. During World War II, he fought against the Japanese—with some help from the United States. After Japan left Vietnam, he proclaimed his country's independence.

ESSENTIAL QUESTION

How have physical barriers in Southeast Asia influenced its history?

The Vietnam War involved the United States and led to bitter divisions in that country. Section 2.4 uses documents to explore the conflict.

INTRODUCE & ENGAGE

Citing Advantages and Disadvantages Pose this situation to students: a small country is fighting for independence against a much more powerful but distant country. Have students form teams and have each team brainstorm the advantages and disadvantages that each side in the conflict would have. After a few minutes of brainstorming, make a list of students' ideas. Revisit the list after reading the lesson for students to clarify their ideas. 0:10 minutes

TEACH

Guided Discussion

1. **Summarize** Who was fighting in the Vietnam War and what were their goals? *(North Vietnam wanted to unite the northern and southern halves of the country under Communist rule. South Vietnam wanted to remain separate, and the United States fought along with it.)*
2. **Analyze Primary Sources** What does President Johnson say would happen if the United States stopped fighting in Vietnam? *(Countries around the world that counted on U.S. protection would lose confidence in the United States.)*
3. **Identify** What does Ho Chi Minh blame for the war? *(U.S. aggression against the people of Vietnam)*

Make Inferences Direct students' attention to the quote from Ho Chi Minh. **ASK:** What does Ho Chi Minh mean by the word *real* when he says "until they have won real independence"? *(Possible response: He is referring to independence from colonial rule or foreign influence in both North and South Vietnam.)* 0:10 minutes

DIFFERENTIATE

Connect to NG

English Language Learners Find Main Idea and Details Have students form groups divisible by three. Assign each group one of the documents. Have native speakers work with each group and start by reading aloud the assigned document, defining and rephrasing as needed. Then have each group use a graphic organizer like the one shown below to identify the important details in their assigned document. Have them determine the main idea based on those details. When finished, each group should state its conclusions to the class.

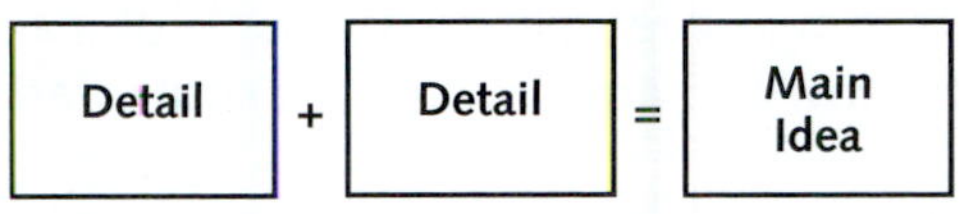

Pre-AP Research the Vietnam War Have students use the **Research Links** to research and create a report that summarizes the Vietnam War in terms of the actions of North and South Vietnam. Students should identify key events in the course of the war, both victories and setbacks, for both parts of Vietnam. Tell them to aim to answer these questions:

- Why did the war end as it did?
- What challenges did both sides face?
- How did they respond to those challenges?

Have students create annotated maps or time lines to represent the key events in the war. Then have them present their visual to the rest of the class. Work together to identify those events that foreshadow the outcome of the war.

ACTIVE OPTIONS

Interactive Whiteboard
GeoActivity

Compare and Contrast Two Wars in Asia Have students complete the Venn diagram in small groups. Suggest that two groups discuss their similarities and differences and try to reach a consensus. 0:15 minutes

On Your Feet

Create a Sequence Chain Divide the class into seven groups and post these dates: 1954, 1959, 1964, 1965, 1967, 1973, 1975. Assign each group a date and have them write a statement about the way in which their assigned year was significant in the Vietnam War. Then have groups stand up in chronological order and recite their statements to construct an oral sequence of events. 0:15 minutes

Performance Assessment

Hold a Roundtable Have teams of students stage a roundtable discussion. Assign one of five topics—ancient times, European colonialism, Indonesia, the Philippines, and Vietnam—to individual students. Tell each student to prepare talking points on his or her assigned topic that address the Essential Question: "How have physical barriers in Southeast Asia influenced its history?" Have the groups take turns staging their roundtable. Go to **myNGconnect.com** for the rubric.

CONSTRUCTED RESPONSE ANSWERS

1. Both reasons reflect the idea that fighting in Vietnam will prevent more conflict.
2. He says peace will come only when the U.S. stops participating in the war.
3. Possible response: Soldiers faced difficult conditions fighting in the jungle terrain of Vietnam.

ONGOING ASSESSMENT

ANSWERS

Steps 1 and 2 Responses will vary.
Step 3 Possible response: Politics, personality, and geography affected North Vietnam's ability to fight and win the war against South Vietnam and the United States.
Step 4 Students' paragraphs may argue that the Vietnamese had a good chance because they were determined and the U.S. was far away. Also, they were adapted to the country and could fight effectively.

CHAPTER 21
Review

VOCABULARY

Use the following vocabulary words in a sentence that shows understanding of each term's meaning.

1. landlocked

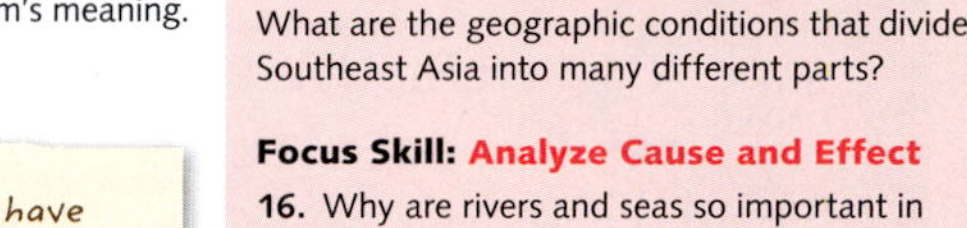

2. subsistence fishing
3. dynamic
4. dormant
5. commerce
6. launch

MAIN IDEAS

7. How does the climate of the two parts of Southeast Asia differ? (Section 1.1)
8. Why are the upper reaches of the region's rivers less populated? (Section 1.2)
9. Where do you think most people live on mainland Malaysia? Why? (Section 1.3)
10. How do Manila and Java show that population in the Philippines and Indonesia tends to cluster? (Section 1.4)
11. What has allowed some species in the region to have gone undiscovered? (Section 1.5)
12. Why did trade play such a great role in the development of island kingdoms? (Section 2.1)
13. Which region of the world practiced colonialism in Southeast Asia and why? (Section 2.2)
14. What struggles did Indonesia and the Philippines face in developing their own culture? (Section 2.3)
15. What were North Vietnam and South Vietnam fighting about? (Section 2.4)

GEOGRAPHY

ANALYZE THE ESSENTIAL QUESTION

What are the geographic conditions that divide Southeast Asia into many different parts?

Focus Skill: Analyze Cause and Effect

16. Why are rivers and seas so important in this region?
17. Why are island countries like Indonesia and the Philippines good for farming?
18. In what way has Laos's lack of a coastline impacted its ability to engage in international trade?

INTERPRET MAPS

19. **Summarize** What is the advantage of the location of the mainland portion of Malaysia?
20. **Make Inferences** Look at the locator map. In what ways has Malaysia's location helped it establish trading partners?

HISTORY

ANALYZE THE ESSENTIAL QUESTION

How have physical barriers in Southeast Asia influenced its history?

Focus Skill: Draw Conclusions

21. Why do you think one great empire never arose to unite all of Southeast Asia?
22. Which other cultural regions do you think had the most influence on Southeast Asia? Explain your reasons.
23. What made it difficult for Europeans to establish complete control of the countries in this region?

INTERPRET MAPS

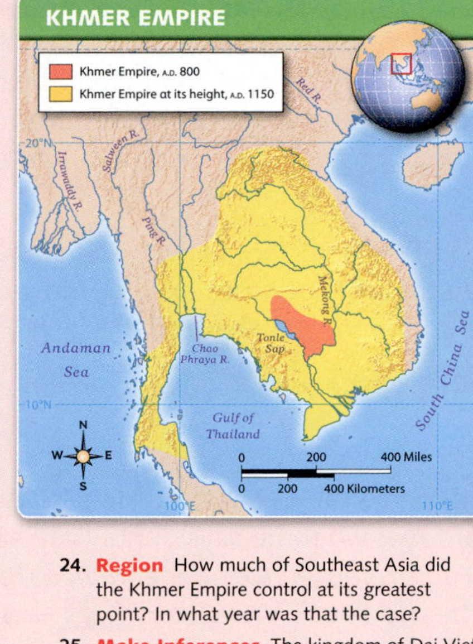

24. **Region** How much of Southeast Asia did the Khmer Empire control at its greatest point? In what year was that the case?
25. **Make Inferences** The kingdom of Dai Viet existed along the South China Sea. What factors would have allowed it to remain independent of the Khmer Empire?

ACTIVE OPTIONS

Synthesize the Essential Questions by completing the activities below.

26. **Write a Press Release** You work for an art museum that is staging an exhibition of art and objects from Angkor Wat. Write a 3- or 4-paragraph press release to announce the exhibition. Include information about how long the exhibition will run and the museum's schedule. Be sure to word the press release in a way that would attract visitors to the exhibition. Use these tips to prepare your press release. **Read your press release aloud with a partner to evaluate each other's ideas.**

Writing Tips
- Make sure you include details that will attract visitors.
- Clearly explain what kinds of objects the exhibition will include.
- Remember to use vivid, appealing language to describe the objects.
- Be sure to give the dates and times of the exhibition.

TECHTREK myNGconnect.com For research links on Southeast Asia

27. **Create a Time Line** Use information in the lessons and research links at **Connect to NG** to gather your facts about five key dates in the history of two countries in the region. Then construct a time line showing the dates for both countries and why they are significant. Illustrate the time line with photos connected to some of the events.

	COUNTRY 1	COUNTRY 2
Event 1		
Event 2		
Event 3		
Event 4		
Event 5		

CHAPTER Review

VOCABULARY ANSWERS

1. All countries in Southeast Asia have some coastline except Laos, making it the region's only landlocked country.
2. Many people in Southeast Asia get just enough food to live on through subsistence fishing.
3. Southeast Asian islands are located in a geographic area that is dynamic, or continuously changing.
4. Even a volcano that is considered dormant may erupt suddenly.
5. The city of Manila in the Philippines has always been a center for commerce, or the trade of goods and services.
6. Fear of communism led the United States to launch a military operation in Vietnam.

MAIN IDEAS ANSWERS

7. Both have a tropical climate. Mainland areas are affected by wet and dry monsoon winds while island countries have precipitation year round.
8. The rivers flow through mountainous areas where it would be difficult to farm, so people are less likely to settle there.
9. Students should infer that most people of mainland Malaysia would live in the lowland areas where the land is fertile and there is ready access to water.
10. Manila has about an eighth of the population of all the Philippines, a large share for one city. Java has the largest share of Indonesia's population, even though the country has thousands of islands.
11. Physical barriers in the region make some areas very remote, which means they have not been explored. When scientists reach these areas, they are likely to find new species.
12. The island kingdoms rose to power as a result of their ability to control trade.
13. Europe practiced colonialism in Southeast Asia starting in the 1800s because European powers wanted to control the spice trade.
14. Influence and control by other cultures made it difficult for Indonesia and the Philippines to establish their own culture.
15. They were at war because communist North Vietnam wanted to overthrow the government of South Vietnam to create a single communist country.

GEOGRAPHY

ANALYZE THE ESSENTIAL QUESTION ANSWERS

16. Rivers allow transportation in mainland countries where mountains are barriers; seas allow transportation to connect the different parts of island countries.

17. Both have fertile volcanic soil as well as warm climates with plentiful rain.

18. Laos has fewer options for shipping because it has no access to a seaport.

INTERPRET MAPS

19. Mainland Malaysia is located along the South China Sea, making it easy to get to China and other parts of East Asia; it has access to the Indian Ocean through the Strait of Malacca; and it is near Indonesia and the Philippines. As a result, it can trade with many different lands.

20. Malaysia is located in the center of Southeast Asia in the middle of trade routes between the Pacific and Indian Oceans.

HISTORY

ANALYZE THE ESSENTIAL QUESTION ANSWERS

21. Possible response: Because the areas in this region are so scattered, because many have physical barriers like mountains, and because the plentiful islands would create areas where people could resist conquest, it would have been difficult to unite these lands under one empire.

22. Possible response: China and India had the most influence in the region because that influence began in ancient times and has been long lasting.

23. Possible response: The areas of the region are so fragmented that it was difficult for European countries to control them.

INTERPRET MAPS

24. At the height of its power in the 1100s, the Khmer Empire controlled almost all of mainland Southeast Asia except the southernmost part of the Malay Peninsula.

25. Possible response: The Khmer Empire was largely land based. It probably did not have much sea power; it probably could not compete with the naval power of the island countries. Its rulers might also have been preoccupied with controlling and defending the lands they had already gained.

ACTIVE OPTIONS

WRITE A PRESS RELEASE

26. Students' press releases should
- include details that will attract visitors;
- explain the kinds of objects the exhibition will include;
- use vivid, appealing language to explain the objects;
- give the dates and times of the exhibition.

CREATE A TIME LINE

27. Students' charts and time lines will vary depending on the countries they choose to focus on. Their time lines should include at least five dates for each of the two countries. Entries should indicate the significance of each listed event.

	COUNTRY 1:	COUNTRY 2:
Event 1		
Event 2		
Event 3		
Event 4		
Event 5		

CHAPTER PLANNER

SECTION SUPPORT

TE Resource Bank

myNGconnect.com

SECTION 1 CULTURE

1.1 Religious Traditions

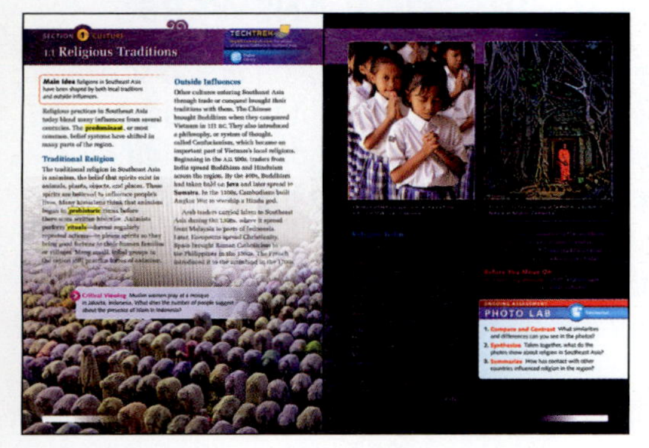

OBJECTIVE Identify religions that have been prominent in Southeast Asia in the past and present.

Reading and Note-Taking
Categorize Religions

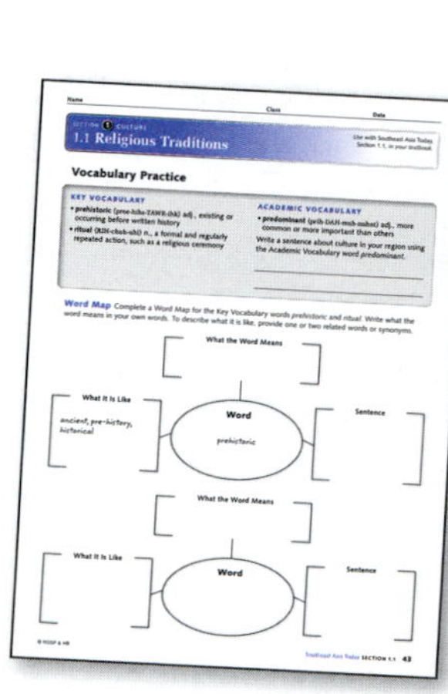

Vocabulary Practice
Word Map

Whiteboard Ready!

GeoActivity
Map Religion in Southeast Asia

SECTION 1 CULTURE

1.2 Thailand Today

OBJECTIVE Analyze how Thailand's culture is influenced by traditional practices and modern life.

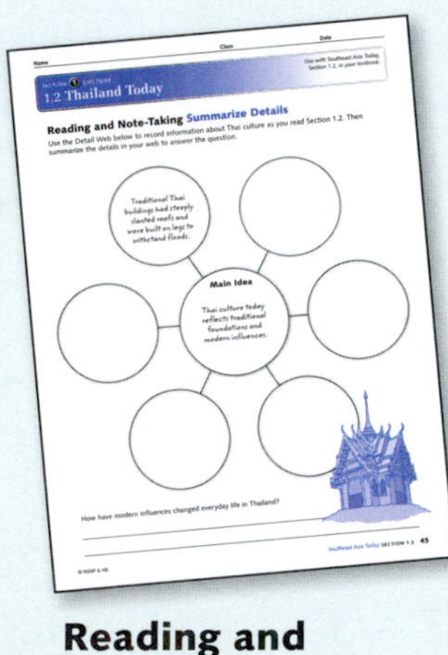

Reading and Note-Taking
Summarize Details

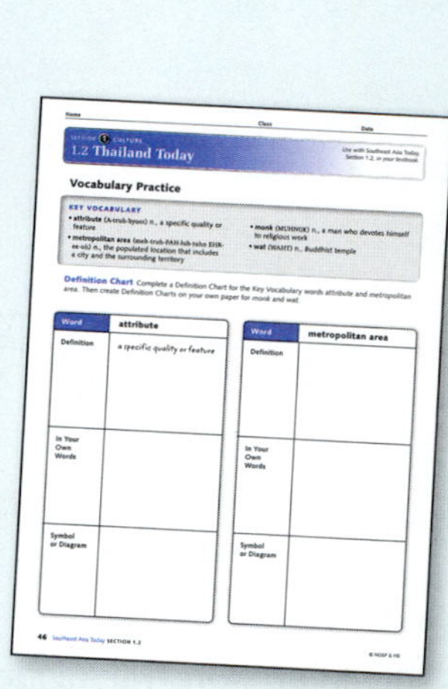

Vocabulary Practice
Definition Chart

Whiteboard Ready!

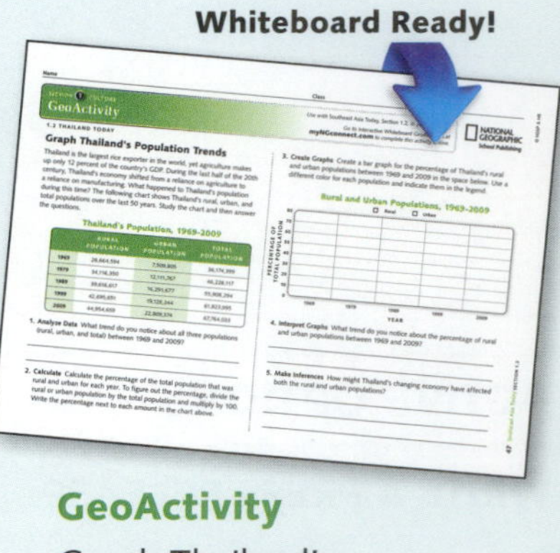

GeoActivity
Graph Thailand's Population Trends

SECTION 1 CULTURE

1.3 Regional Languages

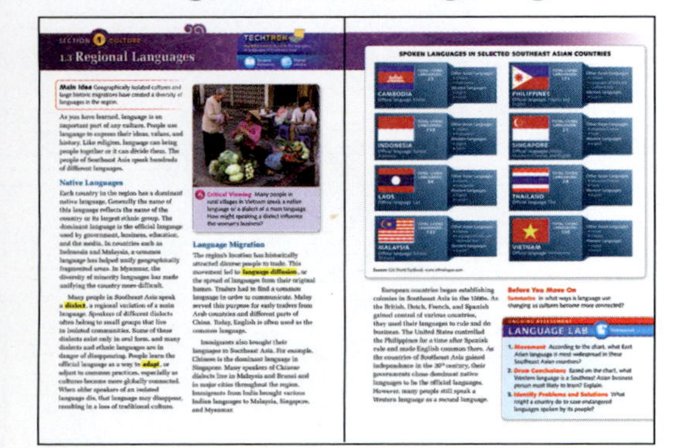

OBJECTIVE Understand the variety of languages spoken in Southeast Asia.

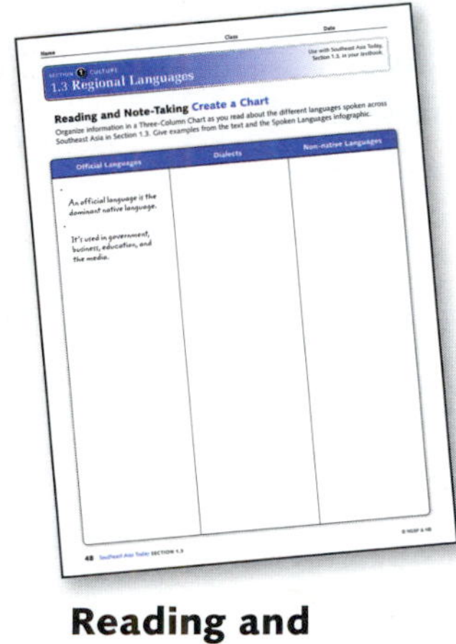

Reading and Note-Taking
Create a Chart

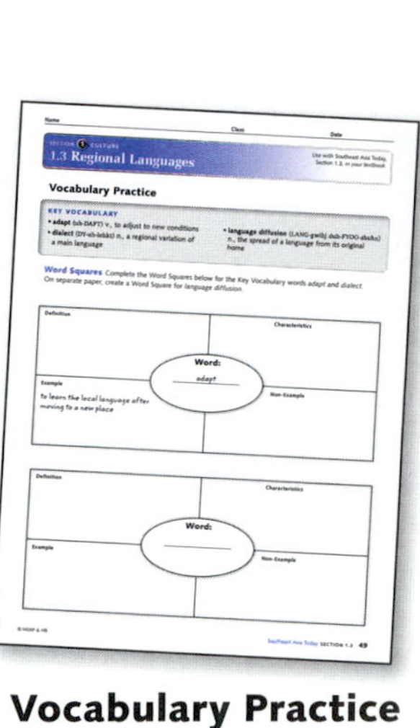

Vocabulary Practice
Word Squares

Whiteboard Ready!

GeoActivity
Analyze Language Relationships

ASSESSMENT

Student Edition
Ongoing Assessment: Photo Lab

Resource Bank and myNGconnect.com
Review and Assessment, Sections 1.1–1.4

ExamView®
Test Generator CD-ROM
Section 1 Quiz in English and Spanish

Student Edition
Ongoing Assessment: Photo Lab

Resource Bank and myNGconnect.com
Review and Assessment, Sections 1.1–1.4

ExamView®
Test Generator CD-ROM
Section 1 Quiz in English and Spanish

Student Edition
Ongoing Assessment: Language Lab

Resource Bank and myNGconnect.com
Review and Assessment, Sections 1.1–1.4

ExamView®
Test Generator CD-ROM
Section 1 Quiz in English and Spanish

TECHTREK myNGconnect.com

Fast Forward!
Core Content Presentations
Teach *Religious Traditions*

Digital Library
NG Photo Gallery, Section 1

Interactive Whiteboard
GeoActivity Map Religion in Southeast Asia

Also Check Out
- Graphic Organizers in **Teacher Resources**
- GeoJournal in **Student eEdition**

Fast Forward!
Core Content Presentations
Teach *Thailand Today*

Digital Library
NG Photo Gallery, Section 1

Connect to NG
Research Links

Maps and Graphs
Interactive Map Tool
Analyze Urbanization in Thailand

Also Check Out
- Graphic Organizers in **Teacher Resources**
- GeoJournal in **Student eEdition**

Fast Forward!
Core Content Presentations
Teach *Regional Languages*

Teacher Resources
Infographic: Spoken Languages in Selected Southeast Asian Countries

Maps and Graphs
Interactive Map Tool
Language Diversity in Southeast Asia

Also Check Out
- Graphic Organizers in **Teacher Resources**
- GeoJournal in **Student eEdition**

CHAPTER PLANNER

SECTION SUPPORT

TE Resource Bank

myNGconnect.com

SECTION 1 CULTURE

1.4 Saving the Elephant

OBJECTIVE Analyze the factors that cause the Asian elephant to be endangered.

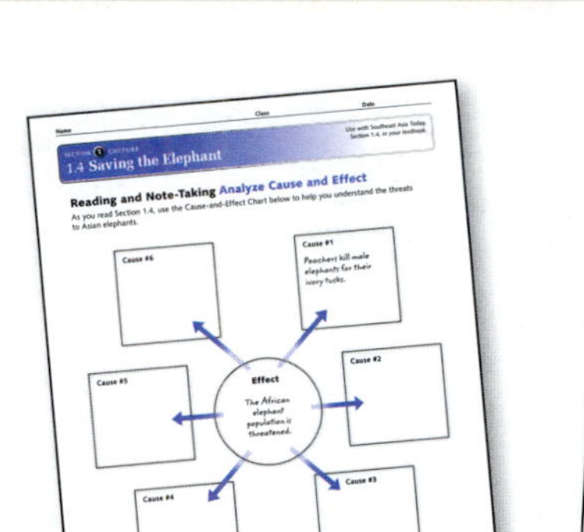

Reading and Note-Taking
Analyze Cause and Effect

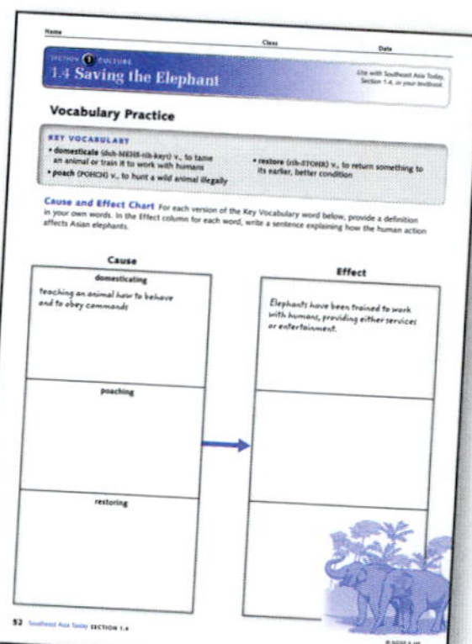

Vocabulary Practice
Cause and Effect Chart

Whiteboard Ready!

GeoActivity
Investigate Endangered Species

SECTION 2 GOVERNMENT & ECONOMICS

2.1 Governing Fragmented Countries

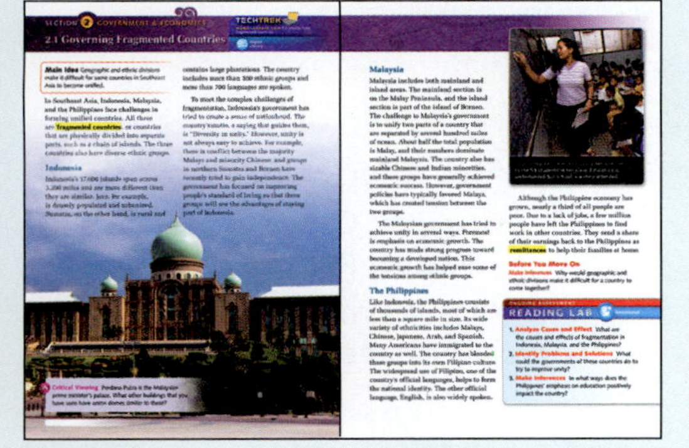

OBJECTIVE Compare the problems of governing Indonesia, Malaysia, and the Philippines.

Reading and Note-Taking
Outline and Take Notes

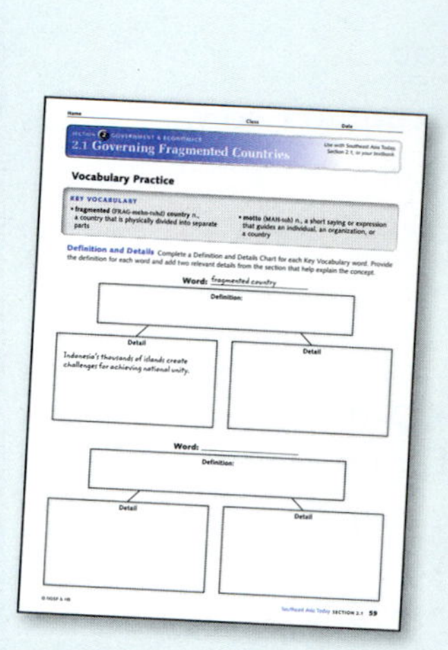

Vocabulary Practice
Definition and Details

Whiteboard Ready!

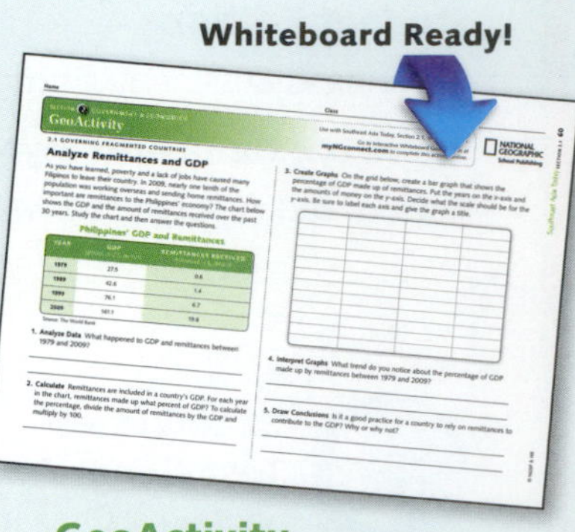

GeoActivity
Analyze Remittances and GDP

SECTION 2 GOVERNMENT & ECONOMICS

2.2 Migration Within Indonesia

OBJECTIVE Explain the process of internal migration and describe its effects.

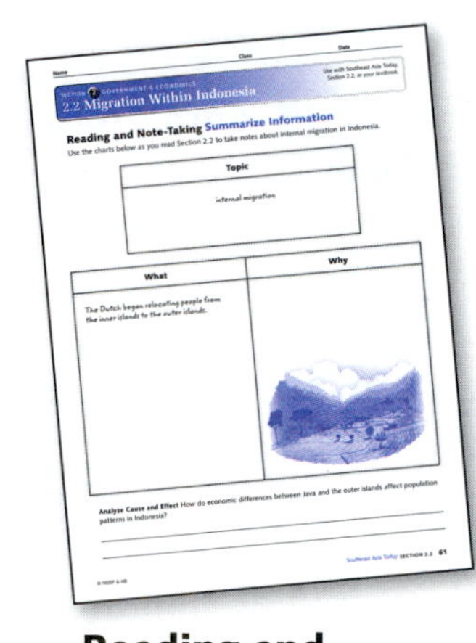

Reading and Note-Taking
Summarize Information

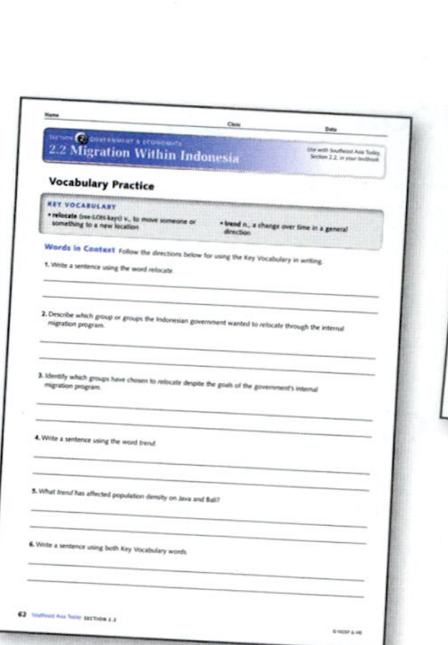

Vocabulary Practice
Words in Context

Whiteboard Ready!

GeoActivity
Evaluate Internal Migration

ASSESSMENT

Student Edition
Ongoing Assessment: Reading Lab

Teacher's Edition
Performance Assessment: Create Multimedia Presentations

Resource Bank and myNGconnect.com
Review and Assessment, Sections 1.1–1.4

ExamView®
Test Generator CD-ROM
Section 1 Quiz in English and Spanish

Student Edition
Ongoing Assessment: Reading Lab

Resource Bank and myNGconnect.com
Review and Assessment, Sections 2.1–2.4

ExamView®
Test Generator CD-ROM
Section 2 Quiz in English and Spanish

Student Edition
Ongoing Assessment: Data Lab

Resource Bank and myNGconnect.com
Review and Assessment, Sections 2.1–2.4

ExamView®
Test Generator CD-ROM
Section 2 Quiz in English and Spanish

TECHTREK myNGconnect.com

Fast Forward!
Core Content Presentations
Teach *Saving the Elephant*

Digital Library
NG Photo Gallery, Section 1

Teacher Resources
Infographic: Asian Elephants by the Numbers

Interactive Whiteboard
GeoActivity Investigate Endangered Species

Also Check Out
- Graphic Organizers in **Teacher Resources**
- GeoJournal in **Student eEdition**

Fast Forward!
Core Content Presentations
Teach *Governing Fragmented Countries*

Digital Library
- GeoVideo: *Introduce Southeast Asia*
- NG Photo Gallery, Section 2

Connect to NG
Research Links

Interactive Whiteboard
GeoActivity Analyze Remittances and GDP

Also Check Out
GeoJournal in **Student eEdition**

Fast Forward!
Core Content Presentations
Teach *Migration Within Indonesia*

Maps and Graphs
- **Interactive Map Tool** Population Density in Indonesia
- Graph: Indonesia: Island Populations

Connect to NG
Research Links

Also Check Out
- NG Photo Gallery in **Digital Library**
- Graphic Organizers in **Teacher Resources**
- GeoJournal in **Student eEdition**

CHAPTER PLANNER

SECTION 2 GLOBAL ISSUES

2.3 Singapore's Growth

OBJECTIVE Analyze how Singapore's free market is related to its economic success.

SECTION SUPPORT

Reading and Note-Taking
Analyze Cause and Effect

Vocabulary Practice
Related Idea Web

Whiteboard Ready!

GeoActivity
Graph Singapore's Economic Rise

SECTION 2 GOVERNMENT & ECONOMICS

2.4 Malaysia and New Media

OBJECTIVE Understand the ways in which new media may change Malaysian society.

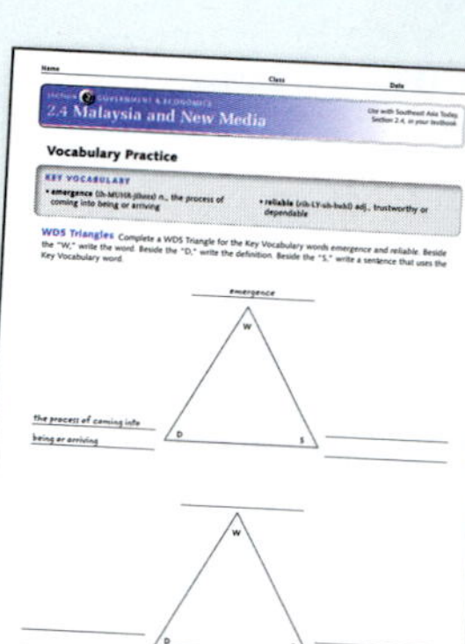

Reading and Note-Taking
Prediction Map

Vocabulary Practice
WDS Triangles

Whiteboard Ready!

GeoActivity
Explore Effects of New Media

CHAPTER ASSESSMENT

INFORMAL ASSESSMENT

TE Resource Bank
myNGconnect.com

Review

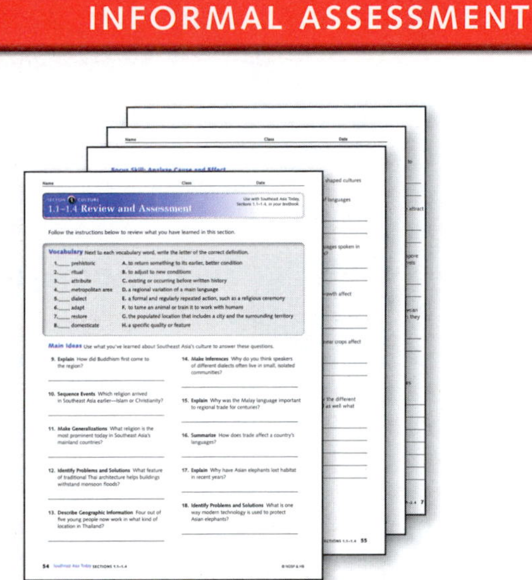

Review and Assessment

Standardized Test Practice

ASSESSMENT

Student Edition
Ongoing Assessment: Writing Lab

Resource Bank and myNGconnect.com
Review and Assessment, Sections 2.1–2.4

ExamView®
Test Generator CD-ROM
Section 2 Quiz in English and Spanish

TECHTREK myNGconnect.com

Fast Forward!
Core Content Presentations
Teach *Singapore's Growth*

Digital Library
- GeoVideo: *Introduce Southeast Asia*
- NG Photo Gallery, Section 2

Connect to NG
Research Links

Interactive Whiteboard
GeoActivity Graph Singapore's Economic Rise

Also Check Out
- Graphic Organizers in **Teacher Resources**
- GeoJournal in **Student eEdition**

Student Edition
Ongoing Assessment: Data Lab

Teacher's Edition
Performance Assessment: Stage a Summit Meeting

Resource Bank and myNGconnect.com
Review and Assessment, Sections 2.1–2.4

ExamView®
Test Generator CD-ROM
Section 2 Quiz in English and Spanish

Fast Forward!
Core Content Presentations
Teach *Malaysia and New Media*

Maps and Graphs
Graph: Southeast Asia: Internet and Cell Phone Access

Interactive Whiteboard
GeoActivity Explore Effects of New Media

Also Check Out
- NG Photo Gallery in **Digital Library**
- GeoJournal in **Student eEdition**

FORMAL ASSESSMENT

Chapter Test A (on level)

Chapter Test B (modified)

ExamView®
Test Generator CD-ROM
Chapter Tests

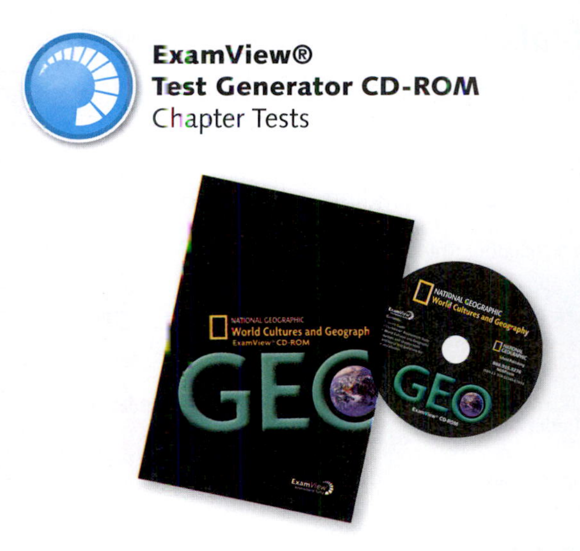

STRATEGIES FOR DIFFERENTIATION

STRIVING READERS

eEdition Audiobook

Strategy 1 • Preview and Set a Purpose

Preview the chapter by reading aloud lesson titles with students. Then ask them to use that information to write a sentence in their own words that begins "I expect to learn about . . ." After reading, ask students to reread their sentences and add the following: "I did learn that . . ."

Use with All Sections

Strategy 2 • Make Lists

Post the title "Three Things I Know about _______ in Southeast Asia" and insert the lesson title in the blank. Ask students to read the lesson and then copy the title and add three sentences about the lesson topic. Students can work individually or in pairs.

Use with All Sections *Students can use subheadings as a guide. For example, Section 1.1 should include one sentence each on "Traditional Religion," "Outside Influences," and "Religion Today."*

Strategy 3 • Use a Cause-and-Effect Web

Display the Cause-and-Effect Web and have students copy it before reading. As they read, they can record the effects of important causes discussed in the chapter.

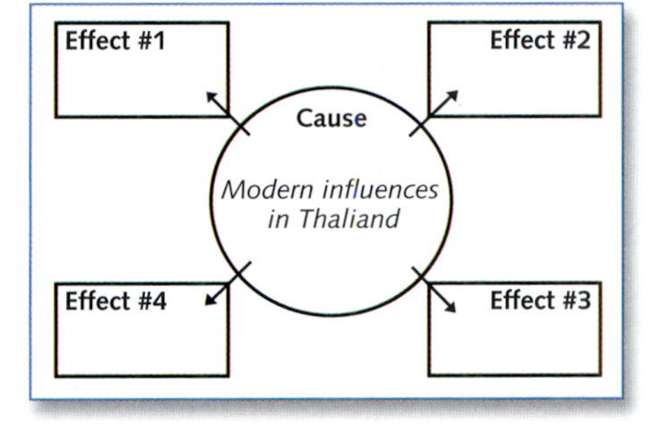

Use with All Sections
Suggested causes might include the following: 1.3, language diffusion; 1.4, declining numbers of Asian elephants; 2.2, internal migration.

Strategy 4 • Create a 3-2-1 Summary

Direct students to complete each sentence frame to build a summary.

3-2-1 Summary

Section	3	2	1
2.1	ways Indonesia is fragmented are ________.	types of divisions in Malaysia are ________.	reason millions have left the Philippines is ________.
2.2	facts about population and size in Indonesia are ________.	kinds of islands in Indonesia are ________.	overcrowded island in Indonesia is ________.
2.4	ways the government of Malaysia controls information are ________.	challenges to government control of information are ________.	Web site that Malaysians go to for information is ________.

Use with Sections 2.1, 2.2, and 2.4

Strategy 5 • Understand Graphics

Remind students that a graphic can be used to display information visually and often uses colors or symbols to represent quantities or ideas. Ask these questions about each graphic in the chapter:

- What is the title? Help students rephrase the title in their own words.
- What colors and symbols are used, and what does each one stand for?
- What message or idea does the graphic communicate?

Use with Sections 1.3, 1.4, 2.2, and 2.4 *In Section 1.4, point out that figures show feet, pounds, and numbers of elephants; in 2.2, point out the legend that shows the number of people each figure represents; in 2.4, read aloud the Graph Tip and point to the part of the graphic that shows more than 100% cell phone usage.*

INCLUSION

eEdition Audiobook

Strategy 1 • Sort Information

Before reading, post a T-Chart with headings related to the lesson, such as those shown below. Provide index cards or sticky notes with details from the lesson. After reading, have students sort the cards or notes into the correct heading in the T-Chart.

1.1 Religious Traditions

OUTSIDE INFLUENCES	RELIGION BROUGHT TO REGION
China	Confucian and Buddhist beliefs
Traders, pilgrims	Hinduism, Buddhism, Islam
Spain, France	Christianity

1.2 Thailand Today

TRADITIONAL FEATURES	MODERN FEATURES
Low, steep roofs	European influence
Built on legs	Uses steel and glass
Made of wood	Combines traditional and modern

Use with All Sections

Strategy 2 • Interpret Graphics

For students who have difficulty with spatial or visual interpretation, provide sentence frames to help them interpret the graphics in this chapter. Suggestions for Section 2 appear below:

2.2 The population of Indonesia's inner islands is _______. The population of the outer islands is (add them) _______. Population is higher on the (inner/outer) islands.

2.3 Between 1960 and 2010, Singapore's annual GDP grew from _______ to _______.

2.4 Internet and cell phone access in Southeast Asia are highest in these three countries: _______.

Use with All Sections

ENGLISH LANGUAGE LEARNERS

eEdition Audiobook

Strategy 1 • Brainstorm to Build Background

Post the word *ritual* and explain its meaning: certain actions done repeatedly, usually as part of a cultural or religious belief. Provide an example students will know, such as the singing of the school song before sporting events. Then write the categories listed below and pair students to brainstorm examples of rituals for each category. Allow pairs to share an example for each category.

religions birthdays sports teams
family holidays

Use with Section 1.1 *This activity can be modified to work with Section 1.2. Teach the word* attribute, *and have students brainstorm attributes of their community or school.*

Strategy 2 • Use a Pair-Share Strategy

After reading, have students work in pairs. The first student reads aloud the Main Idea statement of the lesson and gives one example from the lesson that expands on or illustrates the main idea. Then the second student rereads the Main Idea statement and gives a different example.

Use with All Sections

Strategy 3 • Create Webs for Key Words

Before reading Section 1.2, display the word *architecture* and draw a circle around it. Allow students to volunteer words that they associate with the word *architecture*. Draw spokes for the circle and write the words suggested at the ends of the spokes. Have students write or say sentences that use any of the words in the web. Confirm correctness or guide students in revising. Repeat the procedure for the word *language* before reading Section 1.3.

Use with Sections 1.2 and 1.3 *You may wish to explore other key words depending on students' language proficiency. Possible key words include the following: 2.1,* fragmented; *2.2,* migration; *2.4,* media.

GIFTED & TALENTED

Connect to NG

Strategy 1 • Interview a Community Member

Ask students to prepare a list of questions to use to interview someone in their community who was born in Southeast Asia or has spent some time there. The interviews can be in person or by phone. Allow students to report to the class the results of their interviews.

Use with All Sections

Strategy 2 • Investigate Regional Cuisine

Direct students to choose one country—Vietnam, Thailand, or the Philippines—and research typical foods prepared and eaten in that country. Suggest that they begin by using the **Research Links**. Have students present their information orally or visually, or both.

Use with Sections 1.2, 1.3, and 2.1

PRE-AP

Connect to NG

Strategy 1 • Practice Problem Solving

Allow students to work with partners or in teams to discuss and develop five suggestions for unifying a fragmented country. Suggest that students use the **Research Links** for more information on fragmented countries. Have students report on their suggestions. If time allows, let students discuss which suggestions have the greatest chance of success.

Use with Section 2.1 *This activity can be modified to apply to other lessons in the chapter. For example, for Section 1.4 students might research and suggest other methods of saving Asian elephants in the region.*

Strategy 2 • Form and Support Opinions

Direct students to the **Research Links** and other electronic sources to learn more about why improving the educational level of the workforce could improve the economy of Singapore. Ask them to explain whether the same reasoning would apply to the United States, and why or why not.

Have students explain their reasoning in a "Letter to the Editor" format, as if they are writing to a newspaper in Southeast Asia. You may wish to have students read their own or each other's letters aloud to the class.

Use with Section 2.3

CHAPTER 22

Southeast Asia TODAY

TECHTREK FOR THIS CHAPTER

Student eEdition · Maps and Graphs · Interactive Whiteboard GeoActivities · Digital Library · Connect to NG

Go to **myNGconnect.com** for more on Southeast Asia.

PREVIEW THE CHAPTER

Essential Question How have local traditions and outside influences shaped cultures in Southeast Asia?

SECTION 1 • CULTURE

KEY VOCABULARY

- prehistoric
- ritual
- attribute
- wat
- monk
- metropolitan area
- dialect
- adapt
- language diffusion
- poach
- restore
- domesticate

ACADEMIC VOCABULARY

predominant

TERMS & NAMES

- Java
- Sumatra
- Bali
- Cardamom Mountains

Essential Question How are Southeast Asia's governments trying to unify their countries?

SECTION 2 • GOVERNMENT & ECONOMICS

KEY VOCABULARY

- fragmented country
- remittance
- relocate
- trend
- port
- industrialize
- multinational corporation
- emergence
- reliable

ACADEMIC VOCABULARY

potential

TERMS & NAMES

- Madura
- inner islands
- outer islands
- Malaysia

Tourists and locals use rickshaws and motorbikes to get around in Hanoi, Vietnam.

TECHTREK myNGconnect.com

Fast Forward!

Core Content Presentations
Introduce *Southeast Asia Today*

Digital Library
- GeoVideo: *Introduce Southeast Asia*
- NG Photo Gallery

Maps and Graphs
Interactive Map Tool
Explore Language Diversity in Southeast Asia

Also Check Out
- Graphic Organizers in **Teacher Resources**
- Research Links in **Connect to NG**

INTRODUCE THE CHAPTER

INTRODUCE THE PHOTO

Direct students' attention to the photograph to introduce the issues of modernization, tradition, and urbanization. Point out that, as with most regions, life in Southeast Asia is similar to and different from life in the United States.

ASK: Which features in the photo are similar to those that students might see in a city in North America? Which are different? *(similar: paved street, shops lining it with signs, busy streets; different: streets more full of motorcycles than cars, streets have bicycle-powered rickshaws, streets do not have lane markings)*

COMPARE ACROSS REGIONS

The city of Manila in the Philippines has a population of 11.63 million people while Jakarta in Indonesia has 9.21 million. However, while the Philippines has only two cities with more than one million people, Indonesia has eight. By comparison, the New York-Newark area in the United States has 19.43 million people. **ASK:** How do these population numbers compare to the population in your community? What are some likely benefits of and drawbacks to living in an urban area with a high population? *(benefits: jobs, culture, services; drawbacks: overcrowding, pollution, lack of affordable housing)* Encourage students to use the **Research Links** to discover more about the population in the region.

CONNECT

The city of Hanoi suffered great damage during the Vietnam War. Many ancient monuments were destroyed, and some of these locations remain as historic sites. Its long history and its easy access as a major center for transportation make Hanoi a popular tourist destination.

INTRODUCE THE ESSENTIAL QUESTIONS

SECTION 1 • CULTURE

How have local traditions and outside influences shaped cultures in Southeast Asia?

Four Corner Activity: Methods of Cultural Diffusion This activity helps students explore the different ways in which cultural diffusion takes place. In cultural diffusion, the features of one culture are spread to other cultures and influence those cultures. Post the four signs shown below. **ASK:** Which of these methods of cultural diffusion do you think played the biggest role in the region? Discuss students' choices and the possible positive and negative consequences of each. 0:20 minutes

A Trade Merchants typically carry more than merely trade goods. They can also bring values, practices, and beliefs from their homeland to a new locale. If people in the receiving culture see the traders' culture as valuable, they may adopt features of that culture.

B Migration The movement of peoples has been a constant throughout human history. When two previously unconnected groups come into contact as a result of migration, their cultures often influence each other.

C Conquest When one group conquers another, it brings its own cultural practices and values to its rule. Conquerors often go further and impose their culture. Conquered peoples may accept that culture, resist it, or modify aspects of it.

D Communication Art, ideas, religion, and even cultural practices can be spread through books, newspapers, and magazines; sound recordings; and movies, television, and the Internet. In today's interconnected world, communication may play a greater role than in the past.

If time permits, suggest that students think about other regions they've learned about that have been shaped by outside influences, and which of the above factors brought the greatest influence to the region. For example, culture in North America was strongly influenced by the conquest of Native Americans as well as the influx of immigrants throughout the centuries. After reading Section 1, have students return to this activity to verify or add to their discussion responses. 0:20 minutes

SECTION 2 • GOVERNMENT & ECONOMICS

How are Southeast Asia's governments trying to unify their countries?

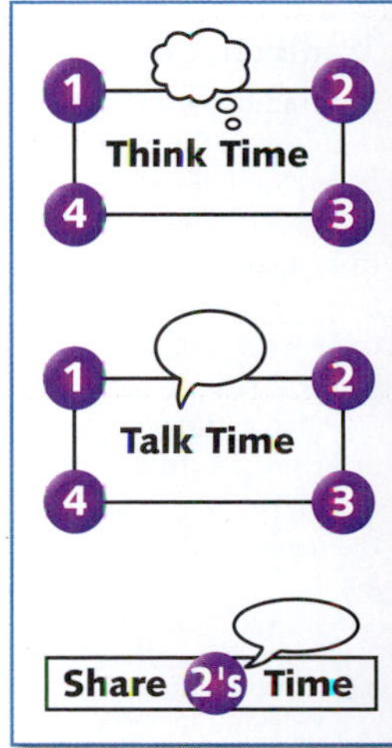

Numbered Heads Activity: Finding Unity One of the challenges facing many communities is diversity. Diversity can enrich a community but may also bring about challenges. Have students form groups of four. Have students individually consider the benefits and challenges that arise from diversity. Have groups choose one benefit and discuss ways their school or community could promote it. Then have them choose one of the challenges and discuss ways their school or community could work to overcome that challenge. Call on one student from each group to present their ideas about promoting the benefit. After all ideas have been presented and discussed, do the same with the groups' suggestions for meeting the challenges. 0:20 minutes

ACTIVE OPTIONS

Interactive Map Tool

Explore Language Diversity In Southeast Asia

PURPOSE Explore the extent of language diversity in Southeast Asia

SET-UP

1. Open the **Interactive Map Tool**, set the "Map Mode" to Topographic, and zoom in on Southeast Asia.
2. Under "Human Systems—Populations & Culture," turn on the Language Diversity layer. Set the transparency level to about 30 percent. Display the legend so that students can interpret the colors.

ACTIVITY

ASK: What do the two countries with the highest language diversity have in common? *(Indonesia and the Philippines are both island countries.)* What can you infer about the relationship between physical geography and language diversity? *(Countries with fragmented physical geography are likely to have greater language diversity.)* As a class, brainstorm some ways a fragmented country's government might address the issue of language diversity and national unity. 0:15 minutes

INTRODUCE CHAPTER VOCABULARY

Teacher Resources

Knowledge-Rating Have students complete a Knowledge-Rating chart for Key Vocabulary. Download the chart from the **Graphic Organizers** and distribute copies to students. See "Best Practices for Active Teaching" for a review of the activity. Have students list words and indicate a knowledge rating for each one. Then have pairs share the definitions they know. Work together as a class to complete the chart.

KNOWLEDGE RATING

KEY VOCABULARY	KNOW IT	NOT SURE	DON'T KNOW	DEFINITION
attribute				
dialect				
industrialize				

SECTION 1 CULTURE

1.1 Religious Traditions

TECHTREK myNGconnect.com For photos of religious traditions in Southeast Asia

Digital Library

Main Idea Religions in Southeast Asia have been shaped by both local traditions and outside influences.

Religious practices in Southeast Asia today blend many influences from several centuries. The **predominant**, or most common, belief systems have shifted in many parts of the region.

Traditional Religion

The traditional religion in Southeast Asia is animism, the belief that spirits exist in animals, plants, objects, and places. These spirits are believed to influence people's lives. Many historians think that animism began in **prehistoric** times before there were written histories. Animists perform **rituals**—formal regularly repeated actions—to please spirits so they bring good fortune to their human families or villages. Many small, tribal groups in the region still practice forms of animism.

Outside Influences

Other cultures entering Southeast Asia through trade or conquest brought their traditions with them. The Chinese brought Buddhism when they conquered Vietnam in 111 B.C. They also introduced a philosophy, or system of thought, called Confucianism, which became an important part of Vietnam's local religions. Beginning in the A.D. 100s, traders from India spread Buddhism and Hinduism across the region. By the 400s, Buddhism had taken hold on **Java** and later spread to **Sumatra**. In the 1100s, Cambodians built Angkor Wat to worship a Hindu god.

Arab traders carried Islam to Southeast Asia during the 1300s, where it spread from Malaysia to parts of Indonesia. Later, Europeans spread Christianity. Spain brought Roman Catholicism to the Philippines in the 1500s. The French introduced it to the mainland in the 1700s.

Critical Viewing Muslim women pray at a mosque in Jakarta, Indonesia. What does the number of people suggest about the presence of Islam in Indonesia?

Islam is widely practiced in Indonesia.

Children attend a prayer service in a Catholic school in Makassar, Indonesia.

Forest roots devour ruins at Ta Prohm, a Buddhist temple at Angkor, Cambodia.

Religion Today

Over the centuries, the predominant religion of a country sometimes shifted depending on the beliefs of the ruling power. Today, Southeast Asia has a mix of religions. Buddhism is most prominent in mainland countries. Around 95 percent of the people in Thailand are Buddhists, and many Buddhist holy days are national holidays. Buddhism is also predominant in Myanmar. Islam is the main religion of Indonesia, which is the most populous Muslim country in the world. In Malaysia, about three out of five people are Muslim. Most people living in the Philippines and East Timor continue to practice Roman Catholicism, introduced by Europeans.

While some religions dominate each country, the region as a whole has religious diversity. For example, **Bali**, an island in Indonesia, is largely Hindu. Islam has many followers in the southern Philippines. Diversity is also found in many of the ancient, local traditions that are still practiced in each country today.

Before You Move On

Monitor Comprehension Which religions were brought to the region by outside cultures? Buddhism, Hinduism, Islam, Christianity

ONGOING ASSESSMENT

PHOTO LAB GeoJournal

1. **Compare and Contrast** What similarities and differences can you see in the photos?
2. **Synthesize** Taken together, what do the photos show about religion in Southeast Asia?
3. **Summarize** How has contact with other countries influenced religion in the region?

PLAN

OBJECTIVE Identify religions that have been prominent in Southeast Asia in the past and present.

CRITICAL THINKING SKILLS FOR SECTION 1.1

- Main Idea
- Monitor Comprehension
- Compare and Contrast
- Synthesize
- Summarize
- Analyze Visuals
- Explain
- Make Inferences
- Create Graphs

PRINT RESOURCES

Teacher's Edition Resource Bank

- Reading and Note-Taking: Categorize Religions
- Vocabulary Practice: Word Map
- **GeoActivity** Map Religion in Southeast Asia

TECHTREK myNGconnect.com

Fast Forward!
Core Content Presentations
Teach *Religious Traditions*

Interactive Whiteboard
GeoActivity Map Religion in Southeast Asia

Digital Library
NG Photo Gallery, Section 1

Also Check Out
- Graphic Organizers in **Teacher Resources**
- GeoJournal in **Student eEdition**

BACKGROUND FOR THE TEACHER

Even within a given religion, there can be diversity of practice. In Indonesia, Muslims in some areas are more traditional than in other areas. Among Christians, too, there are pockets of both Protestants and Roman Catholics. There is also diversity over time. Historically, Java has been home to followers of a type of Islam that is more influenced by local traditions. In recent years, though, more and more Muslims on the island are turning to traditional beliefs.

ESSENTIAL QUESTION

How have local traditions and outside influences shaped cultures in Southeast Asia?

As a lively trading area, Southeast Asia has been influenced by several religions that developed elsewhere. Section 1.1 explores the religious diversity of Southeast Asia and the factors that gave rise to it.

INTRODUCE & ENGAGE

Digital Library

Analyze Visuals Show photos from the **NG Photo Gallery** of people practicing different religions in Southeast Asia. Have students write words and phrases to describe what they see. Remind them to write observations and not judgments, for example, "orange robes" and "hands clasped" rather than "different" or "strange." Post a web like the one shown here, and ask for volunteers to add their observations. 0:15 minutes

Religious Diversity

TEACH

Guided Discussion

1. **Explain** Why might small, tribal groups still practice animism or another ancient, local type of religion? *(Possible response: These groups probably live in greater isolation and have thus maintained traditions. Those in contact with other cultures have been more influenced by those groups.)*
2. **Make Inferences** Review the information about religious diversity within countries in the region. **ASK:** Do you think this diversity would be a strength for the country or a potential problem? Why? *(Possible response: Because religious beliefs are so strongly held, different religious beliefs might come into conflict within a country. On the other hand, the long tradition of different religions in the region could lead people to easily accept that others have different beliefs.)*

Create Graphs Provide students with the data on religious populations in Southeast Asia from the **GeoActivity.** Divide the class into 11 small groups/pairs and assign a country to each one. Have students create a pie chart to show the makeup of religious populations in their assigned country. Have each group show their chart and explain the religious diversity in their country. **ASK:** What might be the effect of one religion being practiced by over 90 percent of a country's population? *(There might be greater unity but also less tolerance of religions other than the majority.)* What might be the effect of greater religious diversity, in which no single religion is dominant? *(The country might be less unified, with greater potential for conflict.)*

Have each group draw a conclusion about the effect of the religious diversity in their country and share it with the class. 0:25 minutes

DIFFERENTIATE

Striving Readers **Categorize Information** Have students work in small groups to organize the information presented in the lesson. Suggest that they make a chart like the one shown here and place a checkmark in the correct boxes.

	BUDDHISM	ISLAM	CHRISTIANITY
Vietnam			
Indonesia			
Philippines			
Thailand			
Myanmar			
Malaysia			
East Timor			

ACTIVE OPTIONS

Interactive Whiteboard

GeoActivity

Map Religion in Southeast Asia Have students complete the activity in small groups to identify the dominant religion in Southeast Asian countries and answer the critical thinking questions. If necessary, review Vietnam's history from the previous chapter. **ASK:** Do you think religion in Vietnam has been more influenced by Chinese conquest or by becoming communist in 1975? *(Possible response: Communism has had more influence on present-day Vietnam because it is more recent.)* 0:20 minutes

EXTENSION Have students circle on their maps those countries that are fragmented. Then have groups work together to draw conclusions about the spread of religion to fragmented countries compared to countries that are on the mainland or are not physically separated. *(Possible responses: Buddhism did not extend beyond mainland Southeast Asia. Islam and Christianity are only dominant in non-mainland countries.)* 0:15 minutes

On Your Feet

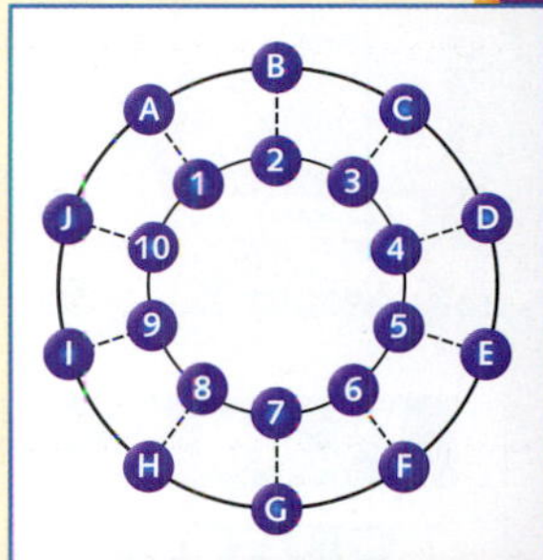

Talking Circles Have students stand in concentric circles facing each other with even numbers of students. Have students take turns quizzing the student standing opposite them about lesson content. If a student cannot answer a question, the student to his or her right should provide a clue to the response. Tell the students to record the questions that were most difficult. Have them discuss those questions as a class. 0:15 minutes

ONGOING ASSESSMENT

PHOTO LAB

ANSWERS

1. Responses will vary. Students might say that the people in all the photos look calm and serious and that in two of the photos they seem to be praying. For differences, they could point to the different clothing, the different settings, and the different numbers of people.
2. Responses will vary, but students should note that the different details in the photos show the variety of religions in the region.
3. Contact with other countries has brought new religions to the region.

SECTION 1 CULTURE

1.2 Thailand Today

TECHTREK myNGconnect.com For photos of modern Thailand

Digital Library

Main Idea Thai culture today reflects traditional foundations and modern influences.

Modern Thai culture includes regional traditions and other global infuences that blend into a unique Thai identity.

Classical Architecture

One **attribute**, or specific quality, of Thai culture is its remarkable architecture. Traditional buildings have steeply slanted roofs designed to shed the heavy monsoon rains. Many are built on legs to keep them high off the ground during the monsoon floods. The most important buildings in Thai architecture are **wats**, or Buddhist temples, influenced by designs from India, the Khmer empire, and China.

Visual Vocabulary A **wat** is a Buddhist temple. Building on Marble Wat (above) started in 1900, during a period of rapid growth in the capital city of Bangkok.

Buddhist Monks

As you have learned, Buddhism is the dominant religion in Thailand. Almost every village has a wat with a community of **monks**, men who devote themselves to religious work. Buddhist monks wear orange or yellow robes, live simply, and focus on religious practices such as meditation and other rituals.

Most young men traditionally became monks for at least three months during one rainy season. However, as more and more young people migrate away from rural communities and attend non-religious schools, young men are making shorter commitments to religious life, or sometimes none at all.

The Chao Phraya River cuts through downtown Bangkok at dusk.

Modern Influences

About four out of five young men and women now work in cities, especially Bangkok's large **metropolitan area**, the populated location that includes the city limits and surrounding area. Many still identify with their villages even though they mostly live and work in cities.

Urban life has also changed people's clothing, food, and entertainment. Most people in Thailand now wear Western-style clothing. Many urban women buy prepared food at local stores on their way home from work instead of cooking. Most homes now have televisions and other modern conveniences.

Traditional: architecture; Modern: fewer young men become monks, young people wear Western clothing and move to urban areas, people use modern technology

Young people in urban areas also turn to the Internet as a source of news, entertainment, and communication.

Before You Move On

Monitor Comprehension What traditional and modern influences can be seen in Thai culture today?

Critical Viewing Suvarnabhumi Airport in Bangkok opened in 2006. Based on the photo, what can you say about the design for the airport? The curved walls and the use of steel and glass suggest a modern design.

ONGOING ASSESSMENT

PHOTO LAB GeoJournal

1. **Make Inferences** What does the large photo show about life in modern Thailand?
2. **Analyze Visuals** What does each photo reveal about Thai culture? Use details from the photos to explain your answer.
3. **Human-Environment Interaction** How was traditional Thai architecture well suited to its environment?

PLAN

OBJECTIVE Analyze how Thailand's culture is influenced by traditional practices and modern life.

CRITICAL THINKING SKILLS FOR SECTION 1.2

- Main Idea
- Monitor Comprehension
- Make Inferences
- Analyze Visuals
- Compare and Contrast
- Draw Conclusions

PRINT RESOURCES

Teacher's Edition Resource Bank

- Reading and Note-Taking: Summarize Details
- Vocabulary Practice: Definition Chart
- **GeoActivity** Graph Thailand's Population Trends

TECHTREK myNGconnect.com

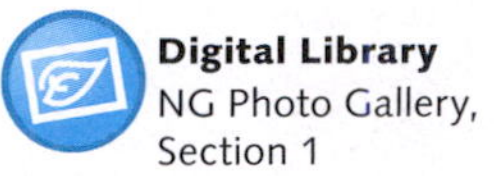

Fast Forward!

Core Content Presentations

Teach *Thailand Today*

Digital Library

NG Photo Gallery, Section 1

Connect to NG

Research Links

Maps and Graphs

Interactive Map Tool

Analyze Urbanization in Thailand

Also Check Out

- Graphic Organizers in **Teacher Resources**
- GeoJournal in **Student eEdition**

BACKGROUND FOR THE TEACHER

The wats of Bangkok reflect traditional styles but are not actually that old. The city only became the capital of Siam, as Thailand was then called, in 1782. The most celebrated wats date mainly from the late 1700s through the mid-1800s. The city's architecture took a decidedly different turn during the reign of King Chulalongkorn (1868–1910), whose public works included structures influenced by French and Italian styles.

ESSENTIAL QUESTION

How have local traditions and outside influences shaped cultures in Southeast Asia?

The people of Southeast Asia have strong local traditions but have also been influenced by interaction with other cultures. Section 1.2 explores how the architecture and society of Thailand have changed over time.

INTRODUCE & ENGAGE

Digital Library

Make Connections Post these descriptions of Thailand's climate: *hot, sunny, heavy rain, strong possibility of flooding*. Tell students to take the role of architects who have to design a house in a climate with the characteristics listed on the board. Allow students time to develop a list of what they think are the best features. Call on volunteers to share their ideas with the class. Then show the photo of a house in Thailand from the **NG Photo Gallery**. Have students identify features of the house that would work well in Thailand's climate. 0:15 **minutes**

TEACH

Guided Discussion

1. **Compare and Contrast** How have the roles of young men changed in Thailand today? *(Since most young men attend non-religious schools and work in cities, it is less common for them to become monks than was traditional.)*
2. **Draw Conclusions** What is the major factor that has contributed to young people in Thailand becoming more modern? *(the shift toward urban life)* **ASK:** Why do you think this shift has had such an influence on young people? *(Young people are usually more interested in technology and other new trends. Young people are now growing up in a global culture.)*

Analyze Visuals Have small groups explore the ways in which a structure is designed to fulfill a certain purpose. Post a chart like the one shown and have each group complete it based on prior knowledge and information in the lesson. Then have groups identify details from the photos that match the purpose of the structure. 0:20 **minutes**

	PURPOSE	NEEDS	DETAILS FROM PHOTO
Wat	A place to worship	Enclosed, quiet, sense of spirituality	Walls, large space around, roof reaching toward sky
Airport	Fast transportation	Fast movement, large crowds	Wide corridors, escalators
Modern office building	Conducting business	Central location, modern conveniences, many people	Tall buildings, many lights, easy transportation

DIFFERENTIATE

Connect to NG

Striving Readers Compare and Contrast Pair students with mixed abilities. Give each pair a T-Chart like the one shown or instruct them to make a T-Chart on a blank piece of paper. Tell the pairs to read through the text and note the features of traditional Thai culture in the left column and modern Thai culture in the right. Pairs can use the charts for review.

Traditional Thai Culture	Modern Thai Culture

Pre-AP Analyze Traditions Have students use the **Research Links** to learn about some aspect of modern Thai culture, such as architecture, religion, transportation, or communication, and determine whether traditional Thai culture is represented. For example, students could start by researching the Mahanakhon building or other high-rise buildings in Bangkok, or students could consider transportation facilities, both modern and traditional.

Have students present their findings to the class. Encourage them to use visual means to share their findings, such as a photo essay or a model of a modern building.

ACTIVE OPTIONS

Interactive Map Tool

Analyze Urbanization in Thailand

PURPOSE Draw conclusions about urbanization patterns

SET-UP

1. Open the **Interactive Map Tool**, set the "Region" to Asia, set the "Map Mode" to Topographic, and zoom in on Thailand.
2. Have students identify the location of Thailand's capital (Bangkok) and the capital of Laos (Vientiane).
3. Under "Environment and Society," turn on the Lights at Night layer. Set the transparency at about 50 percent. Explain that brighter areas are more heavily urbanized.

ACTIVITY

ASK: What might explain the pattern of lights between Bangkok and Vientiane? *(Many people travel between these two capital cities, so the route would become more built up or urbanized.)* Have students work in pairs to investigate the significance of the heavily urbanized area in northern Thailand. *(Chiang Mai is a cultural city that is becoming more modernized.)* **ASK:** What do you predict will happen to the route between Bangkok and Chiang Mai? *(It might become more urbanized as Chiang Mai grows.)* 0:15 **minutes**

NG Photo Gallery

Describe Geographic Information Show the photos of modern Thailand. **ASK:** Based on what you've learned, do these photos reflect traditional Thai culture or an urbanized global culture? Explain. *(Students may observe that the photos look more like modern life in any urban area than like traditional Thai culture.)* 0:15 **minutes**

ONGOING ASSESSMENT

PHOTO LAB

GeoJournal

ANSWERS

1. Life in modern Thailand is urbanized and high-tech.
2. The sloped roofs of Marble Wat reflect traditional Thai architecture. The streamlined shape and the metal and glass shown in the photo of the Bangkok Airport suggest a modern influence. The bright lights and tall buildings in the photo of downtown Bangkok present that city as modernized.
3. Long, steep roofs let rain run off; being built on legs kept buildings above flood waters.

SECTION 1 CULTURE

1.3 Regional Languages

TECHTREK myNGconnect.com For a graphic of languages in Southeast Asia

Student Resources

Main Idea Geographically isolated cultures and large historic migrations have created a diversity of languages in the region.

As you have learned, language is an important part of any culture. People use language to express their ideas, values, and history. Like religion, language can bring people together or it can divide them. The people of Southeast Asia speak hundreds of different languages.

Native Languages

Each country in the region has a dominant native language. Generally the name of this language reflects the name of the country or its largest ethnic group. The dominant language is the official language used by government, business, education, and the media. In countries such as Indonesia and Malaysia, a common language has helped unify geographically fragmented areas. In Myanmar, the diversity of minority languages has made unifying the country more difficult.

Many people in Southeast Asia speak a **dialect**, a regional variation of a main language. Speakers of different dialects often belong to small groups that live in isolated communities. Some of these dialects exist only in oral form, and many dialects and ethnic languages are in danger of disappearing. People learn the official language as a way to **adapt**, or adjust to common practices, especially as cultures become more globally connected. When older speakers of an isolated language die, that language may disappear, resulting in a loss of traditional culture.

Critical Viewing Many people in rural villages in Vietnam speak a native language or a dialect of a main language. How might speaking a dialect influence this woman's business?

Language Migration

The region's location has historically attracted diverse people to trade. This movement led to **language diffusion**, or the spread of languages from their original homes. Traders had to find a common language in order to communicate. Malay served this purpose for early traders from Arab countries and different parts of China. Today, English is often used as the common language.

Immigrants also brought their languages to Southeast Asia. For example, Chinese is the dominant language in Singapore. Many speakers of Chinese dialects live in Malaysia and Brunei and in major cities throughout the region. Immigrants from India brought various Indian languages to Malaysia, Singapore, and Myanmar.

If she knows many dialects, she can communicate with more customers. If her customers don't speak the same dialect, they may not shop there.

SPOKEN LANGUAGES IN SELECTED SOUTHEAST ASIAN COUNTRIES

Country	Official language	Total living languages	Other Asian Languages	Western Languages
CAMBODIA	Khmer	23	Chinese; Vietnamese	English; French
INDONESIA	Bahasa Indonesia	719	Chinese; Vietnamese	English; Dutch
LAOS	Lao	84	Chinese; Vietnamese	English; French
MALAYSIA	Bahasa Malaysia	137	Chinese; Languages of India and Southeast Asia	English
PHILIPPINES	Filipino and English	171	Chinese; Languages of India and Southeast Asia	English
SINGAPORE	Official languages: Malay, Mandarin Chinese, and English	21	Mandarin Chinese; Tamil	English
THAILAND	Thai	74	Mandarin Chinese; Tamil	English
VIETNAM	Vietnamese	106	Khmer; Chinese	English; French

Sources: CIA World Factbook, www.ethnologue.com

European countries began establishing colonies in Southeast Asia in the 1500s. As the British, Dutch, French, and Spanish gained control of various countries, they used their languages to rule and do business. The United States controlled the Philippines for a time after Spanish rule and made English common there. As the countries of Southeast Asia gained independence in the 20th century, their governments chose dominant native languages to be the official languages. However, many people still speak a Western language as a second language.

People are learning official languages as a way of adapting to mainstream culture, and older languages are disappearing.

Before You Move On

Summarize In what ways is language use changing as cultures become more connected?

ONGOING ASSESSMENT

LANGUAGE LAB GeoJournal

1. **Movement** According to the chart, what East Asian language is most widespread in these Southeast Asian countries?
2. **Draw Conclusions** Based on the chart, what Western language is a Southeast Asian business person most likely to learn? Explain.
3. **Identify Problems and Solutions** What might a country do to save endangered languages spoken by its people?

PLAN

OBJECTIVE Understand the variety of languages spoken in Southeast Asia.

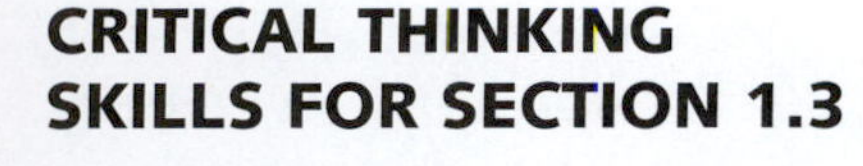

CRITICAL THINKING SKILLS FOR SECTION 1.3

- Main Idea
- Summarize
- Draw Conclusions
- Identify Problems and Solutions
- Analyze Cause and Effect
- Evaluate
- Interpret Graphics

PRINT RESOURCES

Teacher's Edition Resource Bank

- Reading and Note-Taking: Create a Chart
- Vocabulary Practice: Word Squares
- **GeoActivity** Analyze Language Relationships

TECHTREK myNGconnect.com

Fast Forward!

Core Content Presentations

Teach *Regional Languages*

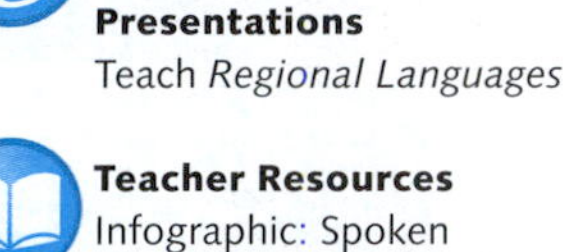

Maps and Graphs

Interactive Map Tool

Language Diversity in Southeast Asia

Teacher Resources

Infographic: Spoken Languages in Selected Southeast Asian Countries

Also Check Out

- Graphic Organizers in **Teacher Resources**
- GeoJournal in **Student eEdition**

BACKGROUND FOR THE TEACHER

The largest language family in the world is the Indo-European, with 2.7 billion speakers, about 46 percent of the world's population. The Sino-Tibetan language family—which is prominent in Southeast Asia—comes next, with 1.3 billion speakers, about 21 percent of all people. A few language families in Southeast Asia have fewer than 1,000 speakers. The smallest is the Bayono-Awbono family of Indonesia, which includes two distinct languages that combine to have only 200 speakers.

ESSENTIAL QUESTION

How have local traditions and outside influences shaped cultures in Southeast Asia?

Language is a key part of culture—one that often changes over time. Section 1.3 examines the diversity of languages in the region and the causes and effects of language diffusion.

INTRODUCE & ENGAGE

Access Prior Knowledge List the top ten most widely spoken languages around the world in alphabetical order: Arabic, Bengali, Chinese, English, German, Hindi, Japanese, Portuguese, Russian, and Spanish. Have students work in pairs or groups to identify what they think are the top five in terms of number of speakers. *(1. Chinese, 2. Spanish, 3. English, 4. Arabic, 5. Hindi)* **ASK:** What might explain why Chinese is the most widely spoken language in the world? *(China's population is high and growing quickly.)* What inference can you make that might explain why no Southeast Asian languages are in the top ten? *(The region is culturally and ethnically diverse, which suggests greater language diversity.)* After reading, have students confirm or correct their inferences. 0:15 **minutes**

TEACH

Teacher Resources

Guided Discussion

1. **Analyze Cause and Effect** What are causes of language diversity in Southeast Asia? *(the number of diverse ethnic groups, the fact that some groups live in isolated areas, migration, the impact of colonial rule, and globalization.)*
2. **Evaluate** Post a T-Chart like the one shown here. Have students work in pairs to brainstorm the advantages and disadvantages of language diffusion. Ask volunteers to add items from their list to the chart. **ASK:** In general, do you think language diffusion is a positive or negative influence in Southeast Asia? Explain. *(Responses will vary, but students might identify the advantages of diversity and the disadvantages of losing tradition.)*

Advantages	Disadvantages

Interpret Graphics Show the Spoken Languages in Selected Southeast Asian Countries infographic from **Charts & Infographics**. Point out the parallel information shown with each flag: country, official language, total living languages, other Asian languages, and Western languages. **ASK:** Which Western language is predominant in Southeast Asia? *(English)* Which country has the highest number of living languages? How can you find this information quickly? *(Indonesia; look at the same space next to each flag)* 0:15 **minutes**

DIFFERENTIATE

English Language Learners **Be the Teacher** Allow English language learners the opportunity to teach their classmates something about their home language. Offer the following suggestions:

- an idiom or special saying that young people use
- a distinct sound or a sound/symbol that does not occur in English
- an expression of an animal sound that is different than it is in English

Allow English language learners to "quiz" their English-speaking peers on what they have learned. **ASK:** What is the value of regularly using your home language? *(so you don't forget how to speak it, and so it does not get lost)*

Gifted & Talented **Debate the Issue** Present two groups with a proposition to debate: *Language diversity divides a country.* Assign one group to the side that agrees with the proposition and the other group to the side that disagrees. Give the two sides time to develop their arguments. Then give them time to debate the issue. Have the teams alternate stating their ideas. When the debate is finished, have the class vote on which position they favor.

ACTIVE OPTIONS

Interactive Map Tool

Language Diversity in Southeast Asia

PURPOSE Explore the extent of language diversity in Southeast Asia

SET-UP

1. Ask students to predict, based on the information in the Spoken Languages in Selected Southeast Asian Countries graphic, the degree to which each country shown has language diversity. Have the class plot each country on a continuum from least diverse (on the left) to most diverse (on the right).
2. Open the **Interactive Map Tool**, set the "Map Mode" to Topographic, and zoom in on Southeast Asia.
3. Under "Human Systems—Populations & Culture," turn on the Language Diversity layer. Set the transparency level to about 30 percent. Display the legend so that students can interpret the colors on the map.

ACTIVITY

Scroll through the region. Have students compare the actual diversity index ranking to their predictions. Discuss as a class any surprises and why the results are surprising. 0:20 **minutes**

On Your Feet

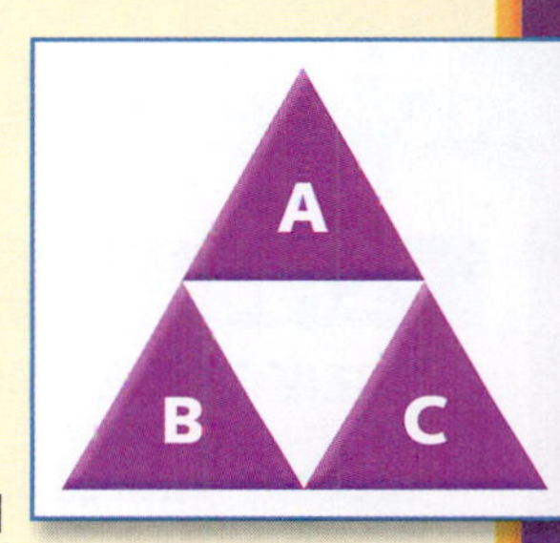

Build a Summary Make cards that say "Causes," "Examples," and "Effects." Place the cards in three different parts of the room and randomly assign students to a group gathered by one of those cards. Have group members work together to make a summary paragraph about their topic with regard to language diversity in the region. Have the groups present their paragraphs. 0:15 **minutes**

ONGOING ASSESSMENT

LANGUAGE LAB

GeoJournal

ANSWERS

1. Chinese; it or a dialect of it is spoken in every country listed on the graphic.
2. English; because it is spoken in every country listed on the graphic
3. Possible responses: The government could collect recordings of people speaking the language and make a dictionary for it. The government could have native speakers teach the language to others.

1.4 Saving the Elephant

TECHTREK myNGconnect.com For photos of elephants in Southeast Asia

Student Resources | Digital Library

Main Idea Some countries in Southeast Asia are working to protect the endangered Asian elephant.

Although smaller than their African cousins, Asian elephants are huge and awe-inspiring creatures. Some have been trained to use their enormous strength for the benefit of people. However, most elephants live in the wild. As the human populations of the region increase, their growing numbers increasingly threaten the Asian elephant population.

Asian Elephants

As recently as 1900, it is estimated 80,000 Asian elephants may have been living in the wild. Today, their population is thought to range from 30,000 to 50,000.

Human behavior, such as **poaching**, or illegal hunting of a wild animal, is one reason for that population loss. People kill male Asian elephants for their ivory tusks. Ivory is highly valued for its beauty and hard texture. An international agreement banned trade in ivory in 1989, but it still continues illegally.

A bigger problem for Asian elephants is their loss of habitat. These huge animals need large areas of rain forest to find food, but humans have cleared much of the land for alternative uses, such as logging and iron mining. Land is also cleared for growing crops such as coffee. Once crops are planted, some elephants wander in from the remaining forests to eat them, and farmers trying to protect their crops sometimes kill those raiding elephants.

Protecting Elephants

Many countries in Southeast Asia have tried to **restore**, or bring back, the wild Asian elephant population. In Cambodia's **Cardamom Mountains**, for example, many conservationists have begun to use modern technology, such as electric fences that run off solar power, to keep elephants confined to protected places. Other methods are more basic. Hammocks hung near crops make the elephants think humans are in the fields, so they stay away.

Critical Viewing Asian elephants look for food in Sumatra. Based on the photo, how would you describe their habitat?

The elephants' habitat is wide-ranging for movement and covered in grass for feeding.

Visual Vocabulary Many elephants are **domesticated**, or trained to work with humans. Domesticated elephants can provide service or entertainment.

The results have been dramatic. The elephant-human interaction that leads to population loss was reduced so much that from 2005 to 2010, no elephants were killed anywhere in Cambodia. These efforts can lead to long-term protection of this endangered species.

ASIAN ELEPHANTS BY THE NUMBERS

11+
Height in feet

12,000+
Weight in pounds

300
Pounds of food (plants, grain) consumed in one day

80,000
Population estimate, wild, beginning of 20th century

30,000
Population estimate, wild, 2008

15,000
Population in captivity (protected), 2006

Sources: World Wildlife Federation, Fauna & Flora International, U.S. Fish & Wildlife Service

Before You Move On

Monitor Comprehension What efforts have been made to protect the wild Asian elephant in Southeast Asia? electric fences, hammocks to keep elephants away

ONGOING ASSESSMENT

READING LAB

GeoJournal

1. **Region** Which country has successfully reduced elephant killings?
2. **Interpret Charts** How has the population of Asian elephants changed over time?
3. **Human-Environment Interaction** What human activities threaten Asian elephants? Why are they a threat?

PLAN

OBJECTIVE Analyze the factors that cause the Asian elephant to be endangered.

CRITICAL THINKING SKILLS FOR SECTION 1.4

- Main Idea
- Monitor Comprehension
- Interpret Charts
- Categorize
- Analyze Data
- Identify Problems and Solutions

PRINT RESOURCES

Teacher's Edition Resource Bank

- Reading and Note-Taking: Analyze Cause and Effect
- Vocabulary Practice: Cause and Effect Chart
- **GeoActivity** Investigate Endangered Species

TECHTREK myNGconnect.com

Fast Forward!
Core Content Presentations
Teach *Saving the Elephant*

Digital Library
NG Photo Gallery, Section 1

Teacher Resources
Infographic: Asian Elephants by the Numbers

Interactive Whiteboard
GeoActivity Investigate Endangered Species

Also Check Out
- Graphic Organizers in **Teacher Resources**
- GeoJournal in **Student eEdition**

BACKGROUND FOR THE TEACHER

Asian elephants have been tamed by humans and used for more than 4,000 years, as shown in clay tablets from the ancient Indus Valley civilization. The people who train and manage elephants are called *oozies* in Myanmar (the name for this occupation in India is *mahout*). Elephants long played an important role in the logging industry, where their great strength could be put to use. Elephants were also used by armies.

ESSENTIAL QUESTION

How have local traditions and outside influences shaped cultures in Southeast Asia?

The largest mammal native to Southeast Asia is the Asian elephant. Section 1.4 explains the dangers to the Asian elephant and efforts to help them.

INTRODUCE & ENGAGE

Digital Library

Categorize Explain that the protection of elephants in Southeast Asia is important not only to save the species but also because elephants are an important part of culture there. Point out the Visual Vocabulary to define *domesticated.* Show photos of Asian elephants from the **NG Photo Gallery.** As students view the photos, have them identify whether the elephants they see are domesticated or in the wild in protected areas. 0:15 minutes

TEACH

Teacher Resources

Guided Discussion

1. **Analyze Data** Show the Asian Elephants by the Numbers graphic, or project it from **Charts & Infographics.** Ask for volunteers to say their own height in feet and inches. **ASK:** How many "students" tall is an Asian elephant? If a student weighs 100 pounds, how many would be needed to match the weight of an Asian elephant? *(120, or around 4 classrooms worth)*
2. **Identify Problems and Solutions** What threat to elephants is not addressed by the efforts described in the lesson? What could be done to reduce that threat? *(These solutions do not address the problem of poaching. Possible solutions are to hire more people to patrol wild areas to protect the elephants or work to educate people on the problem so as to cut the market for illegal ivory.)*

MORE INFORMATION

Loss of Habitat The biggest danger to the elephant population in Southeast Asia is the loss of habitat. This loss is due to the clearing of land for the logging, palm oil, and rubber industries. Humans then settle around these areas. When elephants return to this land that was once theirs, they can trample humans to death and destroy homes and crops. The Wildlife Conservation Society and other activists are working to keep elephants away from land that has been settled by humans and to find illegal loggers.

DIFFERENTIATE

English Language Learners **Explore Vocabulary** Have students work with a native speaker to develop understanding of vocabulary from the lesson. Suggest these options:

- *electric fence, hammock:* Create a sketch using wavy lines to show an "invisible" fence being powered by the sun's rays. Include a simple hammock hung between two trees.
- *domesticate:* Use the Visual Vocabulary to illustrate one way domesticated elephants are used. Offer examples of other domesticated animals such as dogs and horses.

Striving Readers **Create Graphic Organizers** Have students work in pairs. Give each pair a Main Idea diagram like the one shown. Instruct them to write the main idea of the lesson in the top rectangle. Then have them record details from the lesson that support that main idea. Tell them to add more detail boxes if necessary.

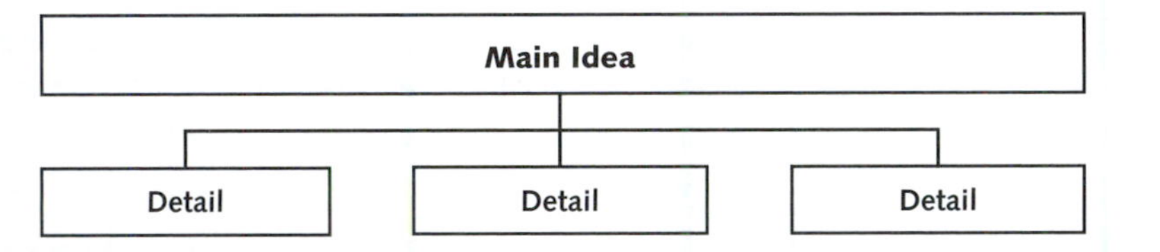

ACTIVE OPTIONS

Interactive Whiteboard

GeoActivity

Investigate Endangered Species After students complete the activity, have them compare notes. First, group together students who researched the same animal. Have them take turns presenting their findings. Then invite other students to state what they learned about the species they studied. As a class, discuss common problems and solutions encountered in protecting these animals. 0:20 minutes

EXTENSION Have students use outline maps to pinpoint specific locations of the species they studied. You may wish to combine these into one large map of protected species in the region. 0:10 minutes

On Your Feet

Finish the Thought Have students stand in two lines facing each other. Tell the students on one side to start a sentence addressing lesson content about Asian elephants. Then have the facing students complete the sentence. If time allows, have students switch roles. 0:15 minutes

Performance Assessment

Create Multimedia Presentations Have teams of four students create a multimedia presentation focused on cultural diversity in Southeast Asia. Assign each student one area to study: religion, language, art, and customs. Tell each group to be sure to include examples from at least two countries in the region. Go to **myNGconnect.com** for the rubric.

ONGOING ASSESSMENT

READING LAB

GeoJournal

ANSWERS

1. Cambodia
2. In 1900, there were about 80,000 of them. Today, there are only about 30,000 to 50,000.
3. Human activities that threaten elephants include poaching; killing them for disturbing crops; and habitat destruction, caused by the clearing of land for commercial use, leaving elephants nowhere to graze.

SECTION 2 GOVERNMENT & ECONOMICS

2.1 Governing Fragmented Countries

TECHTREK myNGconnect.com For photos from fragmented countries

Digital Library

Main Idea Geographic and ethnic divisions make it difficult for some countries in Southeast Asia to become unified.

In Southeast Asia, Indonesia, Malaysia, and the Philippines face challenges in forming unified countries. All three are **fragmented countries**, or countries that are physically divided into separate parts, such as a chain of islands. The three countries also have diverse ethnic groups.

Indonesia

Indonesia's 17,000 islands span across 3,200 miles and are more different than they are similar. Java, for example, is densely populated and urbanized. Sumatra, on the other hand, is rural and contains large plantations. The country includes more than 300 ethnic groups and more than 700 languages are spoken.

To meet the complex challenges of fragmentation, Indonesia's government has tried to create a sense of nationhood. The country's motto, a saying that guides them, is "Diversity in unity." However, unity is not always easy to achieve. For example, there is conflict between the majority Malays and minority Chinese, and groups in northern Sumatra and Borneo have recently tried to gain independence. The government has focused on improving people's standard of living so that these groups will see the advantages of staying part of Indonesia.

Critical Viewing Perdana Putra is the Malaysian prime minister's palace. What other buildings that you have seen have onion domes similar to these?

the Taj Mahal in India; St. Basil's in Russia

Malaysia

Malaysia includes both mainland and island areas. The mainland section is on the Malay Peninsula, and the island section is part of the island of Borneo. The challenge to Malaysia's government is to unify two parts of a country that are separated by several hundred miles of ocean. About half the total population is Malay, and their numbers dominate mainland Malaysia. The country also has sizable Chinese and Indian minorities, and these groups have generally achieved economic success. However, government policies have typically favored Malays, which has created tension between the two groups.

The Malaysian government has tried to achieve unity in several ways. Foremost is emphasis on economic growth. The country has made strong progress toward becoming a developed nation. This economic growth has helped ease some of the tensions among ethnic groups.

The Philippines

Like Indonesia, the Philippines consists of thousands of islands, most of which are less than a square mile in size. Its wide variety of ethnicities includes Malays, Chinese, Japanese, Arab, and Spanish. Many Americans have immigrated to the country as well. The country has blended these groups into its own Filipino culture. The widespread use of Filipino, one of the country's official languages, helps to form the national identity. The other official language, English, is also widely spoken.

Filipino teacher, Leonora Jusay, gives a lesson to the 59 students in her class. Education is underfunded but school is widely attended.

Although the Philippine economy has grown, nearly a third of all people are poor. Due to a lack of jobs, a few million people have left the Philippines to find work in other countries. They send a share of their earnings back to the Philippines as **remittances** to help their families at home.

Before You Move On

Make Inferences Why would geographic and ethnic divisions make it difficult for a country to come together?

Geographic divisions cause difficulties in governing, as each separate part of the country probably lives in a different way. Ethnic divisions can bring about language barriers or ethnic conflict.

ONGOING ASSESSMENT

READING LAB GeoJournal

1. **Analyze Cause and Effect** What are the causes and effects of fragmentation in Indonesia, Malaysia, and the Philippines?
2. **Identify Problems and Solutions** What could the governments of these countries do to try to improve unity?
3. **Make Inferences** In what ways might widespread school attendance in the Philippines' positively impact the country?

PLAN

OBJECTIVE Compare the problems of governing Indonesia, Malaysia, and the Philippines.

CRITICAL THINKING SKILLS FOR SECTION 2.1

- Main Idea
- Make Inferences
- Analyze Cause and Effect
- Identify Problems and Solutions
- Describe Geographic Information
- Summarize
- Evaluate

PRINT RESOURCES

Teacher's Edition Resource Bank

- Reading and Note-Taking: Outline and Take Notes
- Vocabulary Practice: Definition and Details
- **GeoActivity** Analyze Remittances and GDP

TECHTREK myNGconnect.com

Fast Forward!

Core Content Presentations
Teach *Governing Fragmented Countries*

Digital Library
- GeoVideo: *Introduce Southeast Asia*
- NG Photo Gallery, Section 2

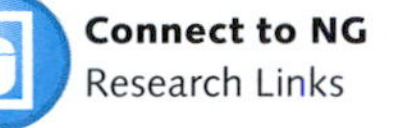

Connect to NG
Research Links

Interactive Whiteboard
GeoActivity Analyze Remittances and GDP

Also Check Out
GeoJournal in **Student eEdition**

BACKGROUND FOR THE TEACHER

Malaysia faces particular challenges to unity, as nine of its states are ruled by hereditary rulers called sultans. (Some are led by governors appointed by the central government.) The head of the national state is the king, who is elected every five years by the nine sultans. One technique the government uses to try to achieve unity is to rotate the kingship regularly through the nine states. Thus, the ruling families know that they each have a stake in maintaining the government.

ESSENTIAL QUESTION

How are Southeast Asia's governments trying to unify their countries?

Governing a country with parts that are isolated from each other poses special problems. Section 2.1 examines the efforts of three governments to unify their fragmented countries.

INTRODUCE & ENGAGE

Digital Library

GeoVideo: *Introduce Southeast Asia* Show the section on fragmented countries in the region and instruct students to pay attention to the features that make a country fragmented. **ASK:** How are Southeast Asian countries divided socially? *(by language, customs, religions)* How are countries divided physically? *(by mountain ranges and coastlines)* 0:15 **minutes**

TEACH

Guided Discussion

1. **Describe Geographic Information** In what specific ways is each country fragmented geographically? *(Indonesia: 17,000 islands over 3,200 miles; Malaysia: mainland and island areas; Philippines: thousands of small islands)*
2. **Summarize** Describe the issues brought about in each country due to its mix of ethnic groups. *(Indonesia: Malays are in the majority and are in conflict with the minority Chinese; Malaysia: policies favor majority Malays, bringing about conflict with minority Indian and Chinese; Philippines: wide variety of ethnicities blend into Filipino culture)*

Evaluate Have students work in pairs to create a chart to organize information about what the governments of fragmented countries have done to promote unity. Then have pairs evaluate the success of these efforts by assigning a score or rank from 1 to 3. Have pairs share their rankings, and discuss why one country's efforts might be more successful than another's. 0:20 **minutes**

	EFFORTS TOWARD UNITY	SCORE
Indonesia		
Malaysia		
the Philippines		

DIFFERENTIATE

Connect to NG

Inclusion Summarize by Matching Provide pairs of students with a set of index cards showing the following words and phrases in mixed order: *17,000 miles; 3,200 ethnic groups; mainland and island areas; thousands of islands; conflict between Malays and Chinese; conflict between Malays and Chinese & Indian groups; tried to create a sense of nationhood; tried to improve people's standard of living; emphasized economic growth; blended ethnic groups into one culture.* Than have students sort the cards into three groups, one each for Indonesia, Malaysia, and the Philippines. Students may use the cards to write a summary or review the lesson.

Pre-AP Report on Other Countries Have students use the **Research Links** to research national unity in one of the other countries in the region. Have them consider the following:

- What geographic, economic, and cultural factors threaten unity or contribute to unity in that country?

Have students write a report that compares unity in the country they studied to unity in one of the countries profiled in the lesson.

ACTIVE OPTIONS

Interactive Whiteboard GeoActivity

Analyze Remittances and GDP After students complete the activity, ask volunteers to share their answers to Question #5. Then post GDP and remittance data for 2009 for all three fragmented countries: Philippines 12.2 percent; Indonesia 1.25 percent; Malaysia 0.56 percent. **ASK:** Given your answer to Question #5, which country is in the best economic position regarding GDP and remittances? *(Responses may vary, but students are likely to point out that Malaysia is in the best economic position because it depends least on remittances.)* 0:20 **minutes**

NG Photo Gallery

Analyze Visuals Show the photos from the **NG Photo Gallery** of Indonesia, Malaysia, and the Philippines. Have students work in pairs to write a caption for each photo that explains how the details in the photo illustrate the country's fragmentation. Then show the photos again, and ask for volunteers to share their captions. 0:15 **minutes**

On Your Feet

Describe a Country Have students form three lines. Give each group one of the three countries profiled in the lesson. Have them describe to the class the degree of unity in that country and the steps the government has taken to try to achieve unity. Tell them to build their descriptions by having each student in the line add one sentence and then go to the back of the line. 0:15 **minutes**

ONGOING ASSESSMENT

READING LAB

GeoJournal

ANSWERS

1. Causes: geographically divided, mix of ethnic and language groups, divided between rural/urban and rich/poor Effects: weaker national identify, conflict between ethnic groups
2. Responses will vary. Possible responses: Supporting economic growth and education for all citizens and discouraging discrimination would support increased unity.
3. A well-educated population can be an important part of decision-making and problem-solving for the country.

2.2 Migration Within Indonesia

TECHTREK myNGconnect.com For photos and a graph of Indonesia's population

Student Resources · Digital Library

Main Idea Efforts to bring unity to Indonesia's islands through relocation have had mixed results.

Indonesia is the fourth most populous country in the world. However, the majority of the people live on just a few of Indonesia's many islands. Living on the remote islands isolates citizens from the greater population, and the country remains fragmented.

Relocation Policy

The Dutch, who had colonized the area, recognized the problem of unifying the vast chain of islands. In the 1800s, they began **relocating**, or moving, individuals and families from Java, **Madura**, and Bali—called the **inner islands**—to the surrounding and less-central islands, the **outer islands**. After Indonesia won independence in 1949, the Indonesian government continued relocating people.

Currently, more than half of Indonesia's people live on the island of Java—an island with a small percentage of the country's total land area. The government hopes that continuing to spread out the Javanese people will help to unify the country. Indonesians native to Java speak the official language, and their presence on the outer islands can help spread the official language to those places where it is infrequently heard. A common language can help unify a fragmented country.

Effects of Internal Migration

So far, however, the practice of moving people among the islands has had unintended results. New arrivals came into conflict with native people, altering their way of life. Modern farming practices clashed with traditional land use and sometimes damaged the environment.

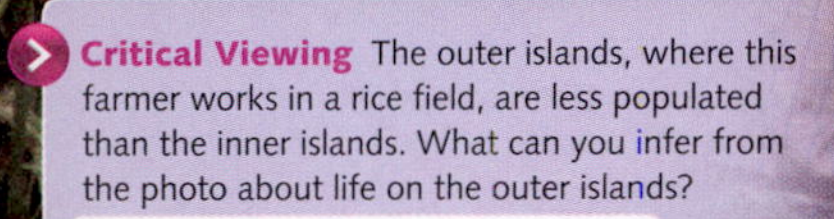

Critical Viewing The outer islands, where this farmer works in a rice field, are less populated than the inner islands. What can you infer from the photo about life on the outer islands?

Life on the outer islands is mostly rural.

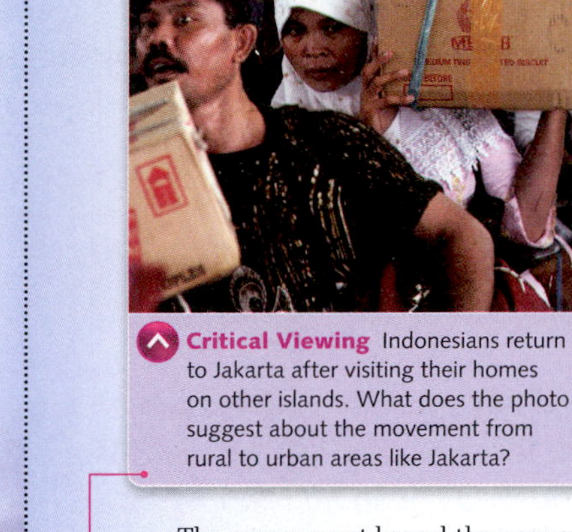

Critical Viewing Indonesians return to Jakarta after visiting their homes on other islands. What does the photo suggest about the movement from rural to urban areas like Jakarta?

The number of people who moved to urban areas is large. Travel back and forth is crowded.

The government hoped the new settlers could build successful farms, but many had trouble supporting themselves. Some ended up abandoning their new homes to return to the inner islands. As a result, Java and Bali remain crowded, making the government's relocation program ineffective. Crowding on Java and Bali grew even worse due to other **trends**, or changes over time. Indonesians living on the outer islands migrated there to flee rural poverty and find work on busier islands. Decades after the program began, Java and Bali are far more densely populated than Indonesia's other islands.

Before You Move On

Summarize How have government policies and economic factors determined migration within Indonesia? Government policies have relocated people from inner islands to outer islands. Lack of economic success drove many migrants back to their homes on inner islands.

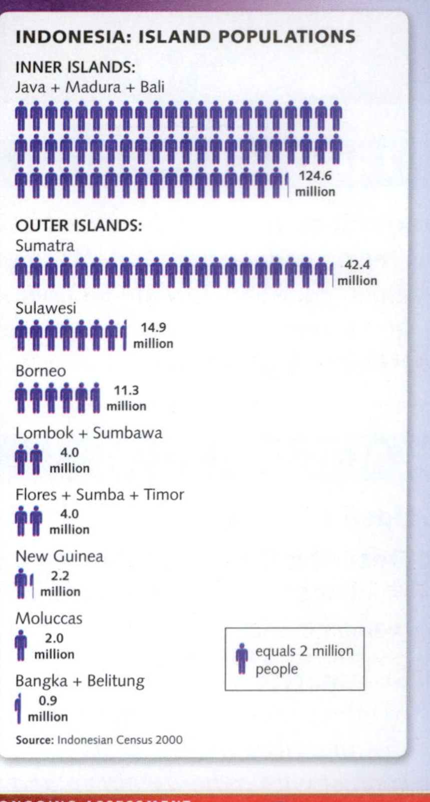

ONGOING ASSESSMENT

DATA LAB

GeoJournal

1. **Interpret Graphs** According to the graph, which is the most populous outer island? Which is the least populous?
2. **Synthesize** When measuring population density, what does higher and lower population density indicate?
3. **Place** Look at the map of Indonesia in Section 1.1 of the previous chapter. Compare the sizes of the islands to the population figures here. Which of the outer islands do you think would have the highest population density? Why?

PLAN

OBJECTIVE Explain the process of internal migration and describe its effects.

CRITICAL THINKING SKILLS FOR SECTION 2.2

- Main Idea
- Summarize
- Interpret Graphs
- Synthesize
- Analyze Causes
- Analyze Effects

PRINT RESOURCES

Teacher's Edition Resource Bank

- Reading and Note-Taking: Summarize Information
- Vocabulary Practice: Words in Context
- **GeoActivity** Evaluate Internal Migration

Fast Forward!

Core Content Presentations

Teach *Migration Within Indonesia*

Maps and Graphs

- **Interactive Map Tool** Population Density in Indonesia
- Graph: Indonesia: Island Populations

Connect to NG

Research Links

Also Check Out

- NG Photo Gallery in **Digital Library**
- Graphic Organizers in **Teacher Resources**
- GeoJournal in **Student eEdition**

BACKGROUND FOR THE TEACHER

The total population of Indonesia is about 243 million. Of this number, approximately 9.1 million live in the capital of Jakarta. The population is growing, and migration away from the country is low. A growing population creates a continuing issue for the government as it tries to create more unity in the geographically and culturally separated country.

ESSENTIAL QUESTION

How are Southeast Asia's governments trying to unify their countries?

Indonesia is a country of many islands and a large population that is concentrated on just a few islands. Section 2.2 discusses how the government has tried to spread out the population to lessen population density and improve unity.

INTRODUCE & ENGAGE

Identify Reasons People Migrate Have students create a graphic organizer like the one shown or give them each a copy of one. Give students a few minutes to write the reasons that people migrate from one region to another. Then call on students to state one of their reasons. Explain that in this lesson, they will learn the causes and effects of a government policy that had people move from one part of the country to another. 0:15 minutes

Why People Migrate

TEACH

Maps and Graphs

Guided Discussion

1. **Analyze Causes** Why did the government decide to move people from Java to outer islands? *(to reduce crowding on Java; to spread people who speak the country's official language to areas where it was not spoken)*
2. **Analyze Effects** What were the unexpected results of this action? *(New settlers came into conflict with native peoples; modern farming methods clashed with traditional land use and disrupted traditional ways of life; the program proved ineffective.)*

Interpret Graphs Direct students to the Indonesia: Island Populations graph or project it from **Graphs**. Explain that in this graph, each figure represents a certain quantity. In this case, a human figure stands for two million people. **ASK:** How many human figures would be used to represent an island population of 10 million? *(five)* Can you determine the population of Java alone from this graph? Why or why not? *(It is not possible to find that population figure from this graph as Java's population is combined with that of Madura and Bali.)* 0:15 minutes

DIFFERENTIATE

Connect to NG

English Language Learners **Use Suffixes** Tell students that the suffix *-tion* as in the word *migration*, means "condition, or state of being." Adding *-tion* turns a verb such as *migrate* into the noun *migration*. List verbs from the lesson and guide students in turning them into nouns by adding the suffix *-tion:*

isolate + tion = isolation	continue + tion = continuation
relocate + tion = relocation	suggest + tion = suggestion
realize + tion = realization	populate + tion = population

Have beginning students work with advanced students or native speakers to write a sentence using each word.

Gifted & Talented **Graph Population Density** Have students calculate the average population density for each of Indonesia's islands—or groups of islands—in the graph. Have them use the **Research Links** or other sources to find the area of each of the islands in square miles. (Remind them to combine the area figures for islands for which population is combined.) Then have them calculate population density by dividing the number of people by the total number of square miles and create a bar graph or other graphic that shows the results. **ASK:** Given the results of the government's relocation efforts, what conclusion can you draw about population density on the islands? *(Responses will vary, but students might point out that population densities will remain high on the island of Java because of the availability of work there.)*

ACTIVE OPTIONS

Interactive Map Tool

Population Density in Indonesia

PURPOSE Compare population density in parts of Indonesia

SET-UP

1. Open the **Interactive Map Tool**, set the "Map Mode" to Topographic, and zoom in on Southeast Asia.
2. Under "Human Systems—Populations & Culture," turn on the Population Density layer. Set the transparency level to about 30 percent.

ACTIVITY

Scroll through the islands of Indonesia, beginning with Sumatra and moving east. Have students assess the population density on the five largest islands using a five-point scale in which 5 is very high and 1 is very low. **ASK:** Based on population alone, which island might be a good place for relocation? *(Responses will vary, but students are likely to identify less populated islands.)* What other factors would be important in determining a relocation policy? *(family structures, availability of resources, opportunity for work)* 0:20 minutes

On Your Feet

Model Migration Have students place chairs into the rough shape of Indonesia and gather by the chairs in rough proportion to the population in the graph. Have a student do a head count and record the numbers on the board. Ask several students in the area of Java, Madura, and Bali to move to other areas. After they do so, have single students from the areas representing the outer islands come to the chairs representing the inner islands. Finally, have two or three of the students who had migrated away from the inner islands return. Do a new head count and compare the results to the original. Discuss how much population density in the inner islands changed. 0:15 minutes

ONGOING ASSESSMENT

DATA LAB

GeoJournal

ANSWERS

1. Sumatra; Bangka + Belitung
2. Population density is the number of people living per square mile or square kilometer of land. Higher density means more crowding; lower density means less crowding.
3. Possible response: Sulawesi, which has the second highest population but is much smaller than Sumatra or the Indonesian part of Borneo

2.3 Singapore's Growth

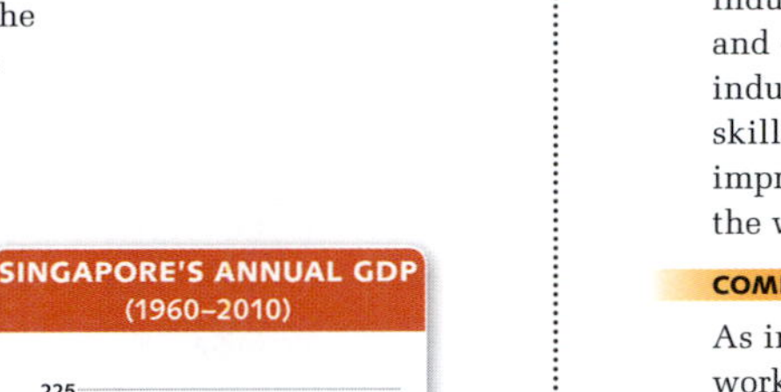

Main Idea Singapore has grown economically due to its geographic location and economic policies.

The British established the modern port of Singapore in the early 1800s in an effort to compete with the Dutch in trade. Located just off the southern tip of the Malay Peninsula, the island has a perfect location near the shipping routes that link the Indian and Pacific oceans. This location gave it the **potential**, or possibility, of becoming a great **port**, a place where ships can exchange cargo. Today, the tiny island country is one of the world's busiest ports—and a strong economic power.

KEY VOCABULARY

port, n., a place where ships can exchange cargo

industrialize, v., to develop manufacturing

multinational corporation, n., large business that has operations in many different countries

ACADEMIC VOCABULARY

potential, n., possibility or promise

Building Success

In 1963, Malaysia gained independence from Britain. Singapore was part of this new country. However, conflict arose between the majority Chinese population of Singapore and the Malays of the rest of Malaysia. In 1965, to reduce the tension, the government offered Singapore its independence, and Singapore took it.

Singapore thrived because of its prime location. It served as the main transit point for sending raw materials such as timber, rubber, rice, and petroleum from Southeast Asia to other parts of the world. Manufactured goods from the United States and Europe came into the port and were shipped to other ports in Southeast Asia. Cars and machinery were shipped into the city from the west to be distributed around the region.

Prime Minister Lee Kuan Yew led Singapore from 1959 to 1990. He emphasized the country's role as an important port and led the drive to **industrialize**, or develop manufacturing. However, the government strictly controlled life in Singapore. Streets were kept clean, and there was very little crime.

Before You Move On

Monitor Comprehension What geographic assets have helped Singapore be part of the global economy? Its prime location as a port city.

SINGAPORE'S ANNUAL GDP (1960–2010)

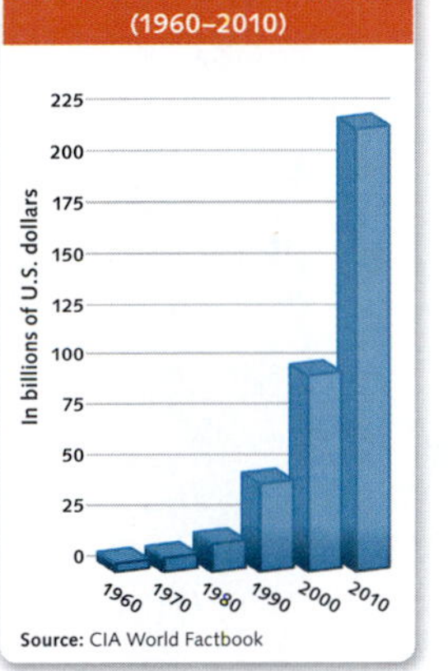

Source: CIA World Factbook

Keeping Pace

From 1965 to 2003, the economic output per person in Singapore grew to more than $24,150. That was more than twice the output of Malaysia. Incomes rose, and unemployment was low. Modern new buildings replaced slums. Singapore became the regional home to many **multinational corporations**.

Singapore's leaders have set goals based largely on economic success. They invest heavily in infrastructure improvements to gain countrywide access to the most current technologies available. To attract foreign investment, Singapore offers low tax rates and other economic incentives. Economic policy focuses on key growth industries such as telecommunications and other technologies. Because these industries rely on educated, highly skilled workers, the country emphasizes improving the level of education in the workforce.

COMPARE ACROSS REGIONS

As in Singapore, a large sector of the workforce in India is highly educated and trained in the newest technologies. Economic policy offers incentives to foreign investors, and many multinational corporations have set up operations there. However, India continues to face many challenges as it works to manage its economic growth. The poverty level is higher and the education level is lower than those of Singapore. India also struggles to improve its infrastructure to keep up with a global economy.

Singapore has been able to participate more fully because of high levels of education, low levels of poverty, and a strong infrastructure. India's struggles in these areas present a challenge to keeping up with globalization.

Critical Viewing This late-night street scene shows the business district in Singapore. Which details in the photo are typical of a prosperous city? office buildings, lots of electric light, traffic on the street

Greater stability has allowed Singapore to take advantage of changes happening elsewhere. When the British handed Hong Kong back to China in 1999, some business owners were concerned that Chinese rule would limit their freedom and prosperity. Singapore welcomed business owners who chose to move their companies there, which boosted economic growth.

Before You Move On

Summarize How have Singapore and India's participation in the global economy differed?

ONGOING ASSESSMENT

WRITING LAB GeoJournal

1. **Make Inferences** Is Singapore's location still important to its economy? Why or why not?
2. **Write Analyses** How has Singapore been affected by globalization? How has its economy been affected by other countries? Go to **Student Resources** for Guided Writing support.

PLAN

OBJECTIVE Analyze how Singapore's free market is related to its economic success.

CRITICAL THINKING SKILLS FOR SECTION 2.3

- Main Idea
- Monitor Comprehension
- Summarize
- Make Inferences
- Analyze Visuals
- Synthesize
- Analyze Causes

PRINT RESOURCES

Teacher's Edition Resource Bank

- Reading and Note-Taking: Analyze Cause and Effect
- Vocabulary Practice: Related Idea Web
- **GeoActivity** Graph Singapore's Economic Rise

Fast Forward!
Core Content Presentations
Teach *Singapore's Growth*

Interactive Whiteboard
GeoActivity Graph Singapore's Economic Rise

Digital Library
- GeoVideo: *Introduce Southeast Asia*
- NG Photo Gallery, Section 2

Also Check Out
- Graphic Organizers in **Teacher Resources**
- GeoJournal in **Student eEdition**

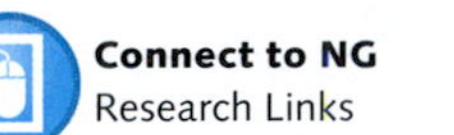

Connect to NG
Research Links

BACKGROUND FOR THE TEACHER

Singapore's economic success is even more impressive given its size. The island nation has only the 100th largest workforce in the world, yet it has the 40th largest economy. (These two rankings do not include the European Union.) One reason for the high ranking is growth: its 2009 GDP (gross domestic product) growth rate of 14.9 percent was second in the world. As a result of having a highly productive small population, Singapore is fifth in the world in per capita output at a staggering $62,200 per person.

ESSENTIAL QUESTION

How are Southeast Asia's governments trying to unify their countries?

Globalization has helped some countries and hurt others. Section 2.3 explains how Singapore has built success on location and a strategic response to globalization.

INTRODUCE & ENGAGE

Digital Library

GeoVideo: *Introduce Southeast Asia* Show the short clip on Singapore. Ask students to explain the description of Singapore's location as "the crossroads of Asia." *(Singapore is a place where many other countries pass through for trade.)*

Analyze Visuals Show the photos of Singapore from the **NG Photo Gallery**. Ask students to point out details that suggest Singapore's economic success. Have them speculate on the factors that have contributed to this success and record them in a T-Chart. Revisit their speculations after reading and have them make any necessary corrections. **0:15 minutes**

Economic Success

Before Reading	After Reading

TEACH

Guided Discussion

1. **Synthesize** How would you describe Singapore's economy as compared to India's? *(Singapore enjoys a greater level of economic success than India, although both countries have used similar techniques to promote economic growth.)*
2. **Analyze Causes** Why do Singapore's leaders invest heavily in infrastructure and education? *(Investing in infrastructure and education has helped boost the economy in the past and made Singapore ready to take advantage of economic opportunities in the region.)*

MORE INFORMATION

Youth Olympic Games In 2007, the International Olympic Committee approved a new event for the first time since 1924, the Youth Olympic Games. The YOG is a competition for dedicated young athletes, ages 14-18, who do not qualify for the Olympics. Singapore was the host of the first summer YOG in August 2010, and the 12-day event was hugely successful. Around 3,600 young athletes from around the world competed in 26 summer sports. Hundreds of teenagers also participated as journalists, ambassadors, or cultural representatives from countries around the world. China is the host of the 2014 Summer YOG.

DIFFERENTIATE

Connect to NG

Striving Readers **Sequence Events** Working in small groups, have students identify the chain of events leading to Singapore's economic success. Remind students to look for dates to indicate the order in which events took place. They should also look for inferences they can make about cause and effect based on the order in which information is presented. Have students present their sequence in a graphic organizer such as the one shown, adding ovals as needed.

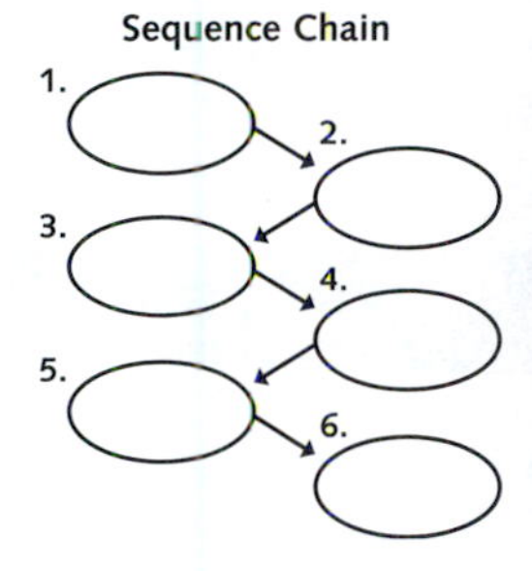

Pre-AP **Compare Countries** Have students use the **Research Links** to find out about the economic performance of one of the other countries in the region in the past two decades. Have them consider the following:

- How does the country's physical geography affect its economy?
- How was the economy performing 20 years ago compared to today?
- What steps has the government taken to promote economic growth?

Have students create charts, graphs, or other visual representations to compare their country with Singapore. Ask students to present their findings to the rest of the class.

ACTIVE OPTIONS

Interactive Whiteboard

GeoActivity

Graph Singapore's Economic Rise Have students complete the graph in small groups to understand the relationship between educational levels and GDP per capita in Singapore. Monitor groups as they plot points on the graph to ensure that they correctly use the *y*-axes. As a class, analyze the data and discuss the graphs. See if there is a class consensus on whether Singapore's strategy of increased emphasis on education has led to greater productivity. **0:15 minutes**

On Your Feet

Explore Contributing Factors Make cards that say "Location," "Manufacturing," "Education," and "Leadership." Place the cards in the four corners of the room and randomly assign students to one of the corners. Have members of each group jointly develop a list of ways that the factor identified on their card has contributed to Singapore's success. Have the groups present their list, with different group members taking turns delivering each point. **0:25 minutes**

ONGOING ASSESSMENT

WRITING LAB

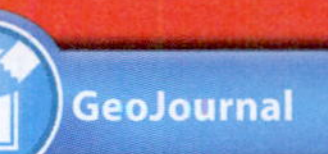

ANSWERS

1. yes, because it is still a busy port due to its location
2. Responses will vary but may include the following points: Singapore has grown as a result of globalization by taking advantage of its location to become a trading center and by serving as a manufacturing and finance center for its region. Its leaders adopted a strategy emphasizing high-technology and information industries A change in the status of Hong Kong led it to welcome the movement of businesses from that city, strengthening its economy.

2.4 Malaysia and New Media

TECHTREK myNGconnect.com For graphs and photos of communication in Malaysia

Maps and Graphs · Digital Library

Main Idea Access to new media sources is changing the strict control on information formerly held by the government in Malaysia.

Since gaining independence in 1963, **Malaysia** has enjoyed prosperity and calm. The government has concentrated on building the economy. However, as in nearby Singapore, Malaysia has limited the freedoms its people can enjoy.

Controlling Information

One of these limits is government control of access to information. Freedom of the press is limited. Newspapers must obtain licenses from the government to operate, and the government can cancel those licenses at any time. Similar laws place restrictions on companies that want to run radio or television stations. Also, the government withholds information from media.

Tough laws punish media outlets that criticize the government. Members of the ruling party own many major newspapers. Even independently owned newspapers usually do not criticize the government.

New Media

The restrictions on the flow of information may be starting to loosen. Nearly two-thirds of Malaysians can now connect to the Internet. New media, like the Internet, give people access to new sources of information that the government may have more difficulty controlling.

Officially the government promises limitless Internet use. However, interference is not uncommon. Many Web site operators focusing on Malaysian news have been arrested multiple times for government criticism. Other journalists have been targeted as well.

Critical Viewing A young woman uses her laptop in a public space in Malaysia. In what ways does the photo show both government control and freedom?

Government control: line of police officers; freedom: Internet access

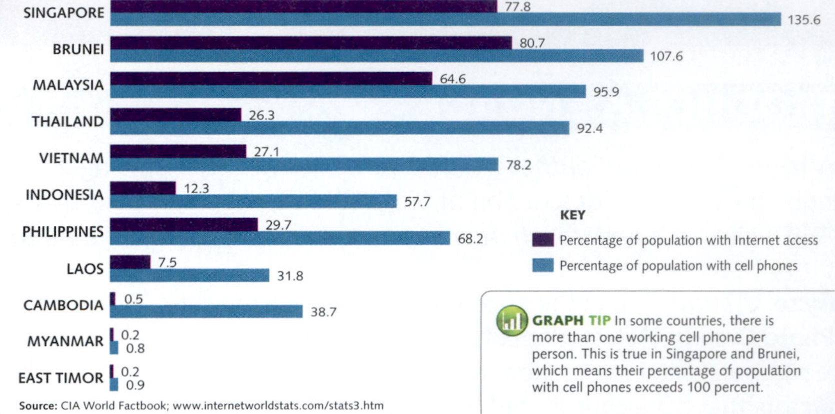

In other instances, police raided Web site headquarters and seized computers in order to find specific people who wrote articles critical of the government.

Many Malaysians believed that until the **emergence**, or arrival, of the Internet the government controlled and manipulated information available to citizens. During elections in 2008, Web sites became the leading source of news for people in the country.

Web sites were outside of government control. They also were completely open to opinions and ideas from anyone in the country. Any citizen could file stories or post videos to the sites. Because of such openness, Web sites may have influenced the 2008 elections. As a result of that vote, the ruling party lost 58 seats in the national legislature. It was the worst result for the party in more than 40 years.

New media technologies give people access to information that is difficult for people in power to control. Poll results found that older citizens trust newspapers and television news and are less likely to embrace additional sources of media. However, among voters in their twenties and thirties, only a small minority trusted traditional media, while more than 60 percent said online news sources were **reliable**, or trustworthy.

Before You Move On

Summarize How has access to new media changed the strict government control of information in Malaysia? The government is not able to keep such a tight control on new media.

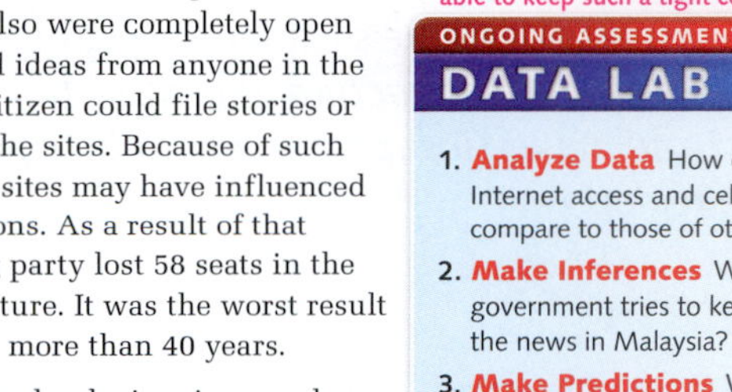

ONGOING ASSESSMENT

DATA LAB

GeoJournal

1. **Analyze Data** How do percentages for Internet access and cell phone use in Malaysia compare to those of other countries in the region?
2. **Make Inferences** Why do you think the government tries to keep such tight control of the news in Malaysia?
3. **Make Predictions** What do the poll results suggest about the way traditional media will be accepted in the future? Why?

PLAN

OBJECTIVE Understand the ways in which new media may change Malaysian society.

CRITICAL THINKING SKILLS FOR SECTION 2.4

- Main Idea
- Summarize
- Analyze Data
- Make Inferences
- Make Predictions
- Explain
- Analyze Visuals
- Interpret Graphs

PRINT RESOURCES

Teacher's Edition Resource Bank

- Reading and Note-Taking: Prediction Map
- Vocabulary Practice: WDS Triangles
- **GeoActivity** Explore Effects of New Media

TECHTREK myNGconnect.com

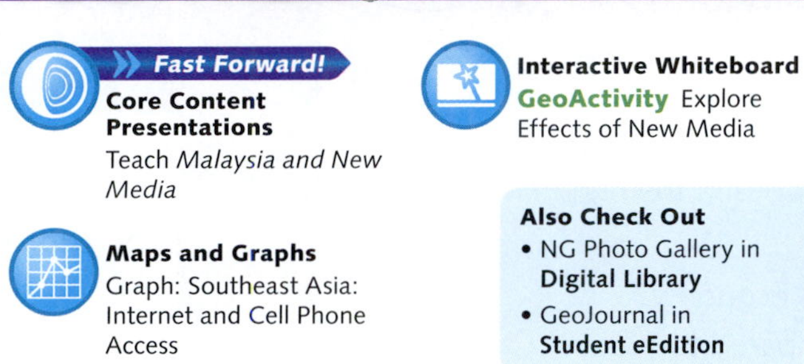

Fast Forward! **Core Content Presentations** Teach *Malaysia and New Media*

Maps and Graphs Graph: Southeast Asia: Internet and Cell Phone Access

Interactive Whiteboard **GeoActivity** Explore Effects of New Media

Also Check Out
- NG Photo Gallery in **Digital Library**
- GeoJournal in **Student eEdition**

BACKGROUND FOR THE TEACHER

Not surprisingly, Indonesia, which has the largest population in the region, ranks highest in terms of the absolute number of people who own cell phones. It is sixth in the world. Vietnam, Thailand, and the Philippines all have more cell phone owners than Malaysia, which ranks 31st in the world. In the number of people with Internet access, Vietnam leads the way in the region and ranks 17th in the world. Indonesia and Thailand are 22nd and 23rd, just ahead of Malaysia, at 26th.

ESSENTIAL QUESTION

How are Southeast Asia's governments trying to unify their countries?

Some governments, like Malaysia's, try to control information to maintain social order. Section 2.4 examines how the spread of new media in Malaysia may be creating governmental policy change.

INTRODUCE & ENGAGE

Brainstorm Benefits Have students brainstorm the benefits of cell phones, Internet access, and other electronic media. Tell them to consider the ways in which these technologies enable people to access information and communicate with one another. After students have had a few minutes to develop their ideas, call on volunteers to state one benefit they identified. Discuss students' responses as a class. Inform students that in this lesson, they will learn how these new media are having an effect on political life in a country in the region. 0:15 minutes

TEACH

Maps and Graphs

Guided Discussion

1. **Explain** What steps has Malaysia's government taken to control information? *(requiring newspapers and radio and television stations to obtain licenses, which it can cancel; withholding information from the media; punishing media outlets that criticize the government)*

2. **Analyze Visuals** How does the photograph of the woman with her laptop support the idea that the new media are changing news coverage in Malaysia? *(The woman could report on what the police are doing as they are doing it, with her statements being instantly available to anyone with Internet access.)*

Interpret Graphs Direct students to the Southeast Asia: Internet and Cell Phone Access graph, or project it from **Graphs**. Ask volunteers to explain what the purple and blue bars represent. *(The purple bar is the percentage of people in the country with Internet access. The blue bar is the percentage with cell phones.)* **ASK:** How can the percentage of people who own cell phones be over 100 percent? *(Some people may own more than one cell phone. There are more cell phones in use in the country than there are people.)* Why could the other measure not have a value over 100 percent? *(Internet access is an either-or proposition: you either have it or you do not.)* 0:15 minutes

DIFFERENTIATE

Inclusion **Rephrase Information** Pair mixed ability students to interpret the visual information presented in the graph. Provide the following sentence frames for students to complete to help them understand visual data:

- The country with the highest level of Internet/cell phone access is _______, with _______ percent.
- _______ has a low level of Internet access but a high level of cell phone usage.
- All Southeast Asian countries have a higher level of _______ access than _______ access.

Encourage students to compose their own sentences interpreting the graph. Then have them work with a partner to check for correctness.

Pre-AP **Look Ahead** Have students look at the sequence of events described in the lesson. Tell them to take the role of a journalist from another country reporting on Malaysia. Have them write and deliver a report that answers these questions:

- How did the government control media in the past?
- How did the government react to new media actions in 2008?
- What prediction can you make about the way the government will treat the new media in Malaysia in the future? Why?

ACTIVE OPTIONS

Interactive Whiteboard
GeoActivity

Explore Effects of New Media Have students complete the activity in small groups to explore the impact of new media on political dissent. Allow groups to compare their responses with other groups and make corrections where necessary. 0:15 minutes

On Your Feet

Sound Off on Media
Have students form groups of equal numbers, with each group standing as shown in the diagram. The two circles should face each other. Tell them to take turns quizzing each other on media and politics in Malaysia. 0:15 minutes

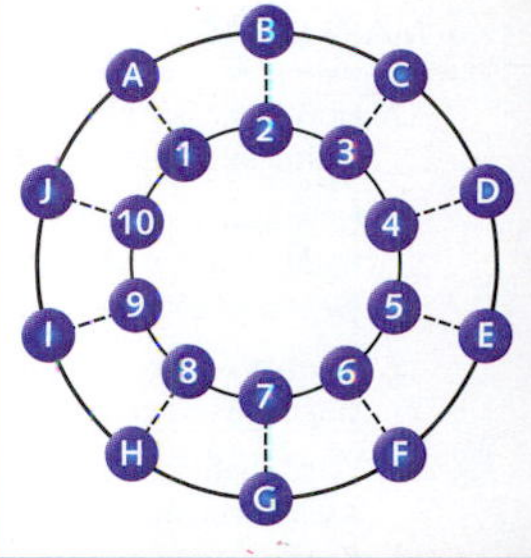

Performance Assessment

Stage a Summit Meeting Divide the class into five groups and assign four groups one of these countries: Indonesia, Malaysia, the Philippines, or Singapore. The fifth will be journalists. Tell the groups that they represent the delegation that the government of their country is sending to a regional summit meeting that will discuss economic development and national unity. Have each group collaborate on a talk to give at the meeting that outlines their country's challenges in these two areas and the steps being taken to meet them. The journalist group should ask each government group questions about its policies, goals, and progress. Go to **myNGconnect.com** for the rubric.

ONGOING ASSESSMENT

DATA LAB

GeoJournal

ANSWERS

1. Malaysia ranks third highest in Internet access and in cell phone use.
2. Possible responses: to prevent criticism; so that the people currently in power will not have their power threatened by opposition groups
3. Since a high percentage of young people do not trust traditional media much now, there are likely to be fewer and fewer people in the country that trust traditional media in the future.

VOCABULARY

For each vocabulary word, write one sentence that explains its meaning and relates it to the content of the chapter.

1. prehistoric

Many religions in Southeast Asia are prehistoric, meaning they existed before history was written down.

2. metropolitan area
3. dialect
4. domesticate
5. relocate
6. multinational corporation

MAIN IDEAS

7. How has colonial history shaped religion in the region? (Section 1.1)
8. How does modern Thailand reflect both tradition and external influences? (Section 1.2)
9. In what way has globalization changed language use in Southeast Asia? (Section 1.3)
10. What methods can help protect the wild Asian elephant population in Southeast Asia? (Section 1.4)
11. How has fragmentation been a problem for Indonesia, Malaysia, and the Philippines? (Section 2.1)
12. What have been the results of Indonesia's policy of moving people to new areas? (Section 2.2)
13. How has Singapore built its economy? (Sections 2.3)
14. How have new media changed politics in Malaysia? (Section 2.4)

CULTURE

ANALYZE THE ESSENTIAL QUESTION

How have local traditions and outside influences shaped cultures in Southeast Asia?

Focus Skill: Make Generalizations

15. How have outside influences increased the diversity of Southeast Asia?
16. What difficulties do the countries of Southeast Asia face in trying to maintain traditional culture in the modern world?

INTERPRET TABLES

PERCENTAGES OF ETHNIC GROUPS IN SELECTED SOUTHEAST ASIAN COUNTRIES

Country		
Indonesia	Javanese: 41% Sundanese: 15% Madurese: 3%	Minangkabau: 3% Other: 38.4%
Laos	Lao: 55% Khmou: 11%	Hmong: 8% Other: 26%
Malaysia	Malay: 50% Chinese: 24%	Indian: 7% Other: 19%
Philippines	Tagalog: 28% Cebuano: 13% Ilocano: 9%	Bisaya/Binisaya: 8% Hiligaynon Ilonggo: 8% Other: 35%
Singapore	Chinese: 77% Malay: 14%	Indian: 8% Other: 1%
Thailand	Thai: 75% Chinese: 14%	Other: 11%
Vietnam	Kinh (Viet): 86%	Other : 14%

Source: CIA World Factbook

17. **Analyze Data** Which three countries have a single dominant ethnic group? What is the group in each case?
18. **Make Generalizations** Would you expect countries with one dominant national group to have a single official national language? Why or why not?

GOVERNMENT & ECONOMICS

ANALYZE THE ESSENTIAL QUESTION

How are Southeast Asia's governments trying to unify their countries?

Focus Skill: Summarize

19. What economic and political concerns led Indonesia to adopt a policy of moving people from the inner islands to the outer islands?
20. Why did Singapore split from Malaysia and become independent?
21. How is the policy of Malaysia's government to limit freedom of the press related to the issue of fragmentation?

INTERPRET MAPS

22. **Interpret Maps** Why is it appropriate to call the Philippines a fragmented country?
23. **Draw Conclusions** What transportation improvements might the government invest in to help unify the country? Explain.

ACTIVE OPTIONS

Synthesize the Essential Questions by completing the activities below.

24. **Write a Feature Article** Introduce a friend to Southeast Asia today by writing a feature article that describes one country in the region. In your article, highlight the ways religious practices and regional languages shape culture in that country. Describe the modern developments and traditional practices that impact the daily lives of people who live there. **Share your article with your friend.**

Writing Tips
- Provide a clear, concise introduction of your country.
- Keep your paragraphs tightly focused on one topic.
- Include details about the country's religions, languages, and form of government.
- Provide smooth transitions between paragraphs.
- In your conclusion, include a paragraph that summarizes your article.

TECHTREK myNGconnect.com For research links on Southeast Asia today

25. **Create Graphs** Use the research links at **Connect to NG** to make a bar graph showing the per capita gross domestic product (GDP) of Cambodia, Malaysia, and the Philippines. Write a paragraph explaining which countries' economies might be affected by fragmentation. Use the example below as a guide for your graph.

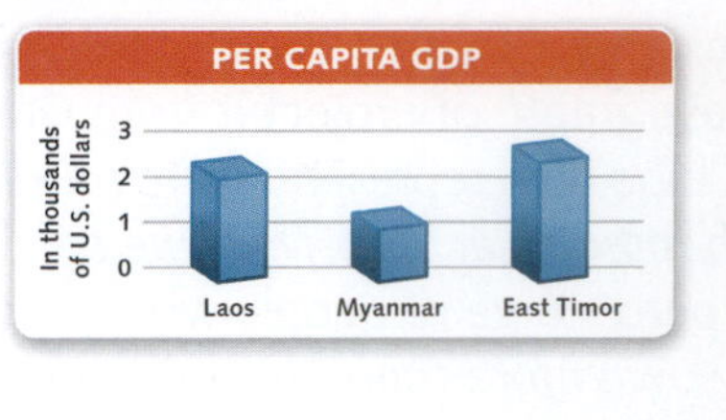

CHAPTER Review

VOCABULARY ANSWERS

1. *Many religions in Southeast Asia are prehistoric, meaning they existed before history was written down.*
2. The metropolitan area around Bangkok is heavily populated by people who have migrated from rural areas.
3. Many of the languages in Southeast Asia are dialects, or local variations on major languages.
4. Animals such as elephants can be trained to work with humans, or domesticated.
5. Relocating from one place to another can be exciting and challenging.
6. A multinational corporation may bring jobs to many different countries because of its various business operations.

MAIN IDEAS ANSWERS

7. Spanish, Portuguese, and French colonial control of the Philippines, East Timor, and Vietnam resulted in the spread of Roman Catholicism to those areas.
8. Traditional architecture exists among more modern buildings. People move to urban areas for work but still identify with their villages.
9. An example of the impact of globalization on language in the region is the spread of English in modern times, as English is a language of business today.
10. Conservationists are using different methods to protect elephants including electric fences to keep elephants in protected areas and hammocks hung in fields to encourage elephants to stay away.
11. At times, members of the majority Malay population have attacked the Chinese minority. Also, groups in northern Sumatra and in Borneo have recently tried to gain independence.
12. The policy has not been a complete success, and many of the people have moved back to the inner islands.
13. Singapore has built its economy by investing in infrastructure, encouraging foreign investment, focusing on growing industries, and promoting education.
14. New media have changed politics in Malaysia by functioning outside of government control and providing information from independent sources.

CULTURE

ANALYZE THE ESSENTIAL QUESTION ANSWERS

15. The influence of nearby areas like India and China, trade with Muslim peoples, and European colonialism have led to the spread of religion and language. These factors, along with the movement of peoples, have made the different countries of Southeast Asia diverse.

16. Possible response: Increasing globalization and the speed of communications probably put a great deal of pressure on the people of Southeast Asia to adopt modern, Western ways, which makes it more difficult for them to maintain traditional cultures.

INTERPRET TABLES

17. Singapore (Chinese); Thailand (Thai); Vietnam (Kinh or Viet)

18. Possible response: They probably have a single national language because the dominant ethnic group forms such a large share of the population that it probably dominates the government as well.

GOVERNMENT & ECONOMICS

ANALYZE THE ESSENTIAL QUESTION ANSWERS

19. The government of Indonesia adopted the policy of moving people to the outer islands because it wanted to reduce overcrowding in the inner islands and sprinkle people from the main ethnic group to more far-flung islands.

20. As a result of conflict between the Chinese of Singapore and the Malays of the rest of Malaysia in newly-independent Malaysia, the government offered Singapore independence, and Singapore took it.

21. Possible response: A fragmented country is harder to govern, so limiting freedom of the press might help the government maintain its control.

INTERPRET MAPS

22. The Philippines can be called fragmented because it is composed of many islands. It also has ethnic diversity and social and economic differences that deeply divide the country.

23. Possible response: The government might reduce divisions by investing in a water transportation system to provide easy travel between islands.

ACTIVE OPTIONS

WRITE A FEATURE ARTICLE

24. Feature articles should include
- a clear, concise introduction;
- focused paragraphs;
- details about religions, languages, and government;
- smooth transitions between paragraphs;
- a summary conclusion paragraph.

CREATE GRAPHS

25. See the graph at right for relevant information. Have students submit their sources as well as the finished product. In their paragraphs, they might identify the countries with the highest GDP per capita as those least likely to suffer from economic unrest.

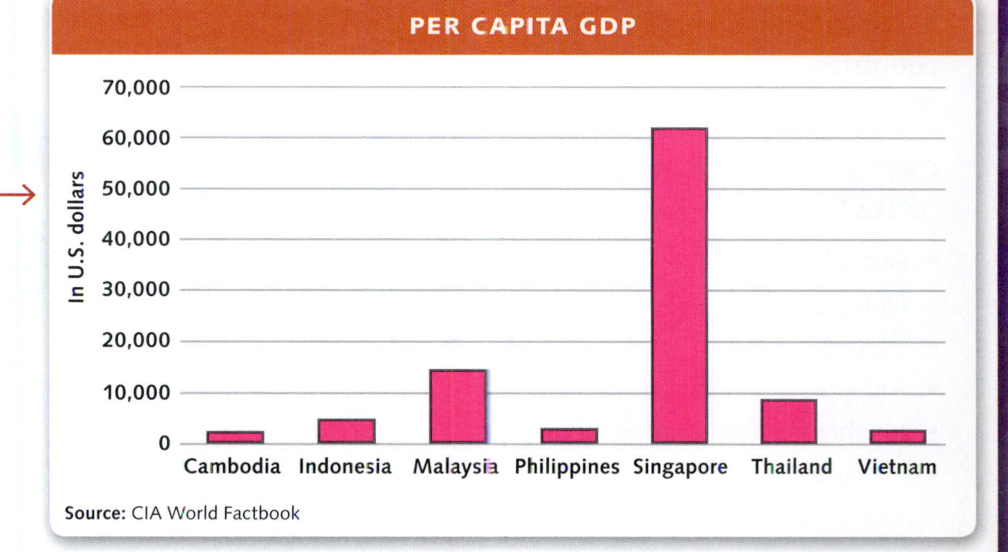

Source: CIA World Factbook

Endangered Species

TECHTREK myNGconnect.com For an online graph and research links on endangered species

Maps and Graphs | Connect to NG

Endangered species of both plants and animals can be found in every region of the world. A species is endangered when it runs the risk of becoming extinct, or disappearing completely from the world. Over time, various species have not survived. In fact, historically there have been five mass exinctions, in which a large number of existing species died out.

Today, the rate of extinction for plant and animal species has become hundreds of times faster than what scientists have observed through the fossil record. Many scientists believe the earth is currently in the midst of a sixth mass extinction. Those same scientists believe the main cause to be the destruction of habitat.

Compare Ranges of Endangered Big Cats

- Asiatic Lion
- Cheetah
- Iberian Lynx
- Jaguar
- Tiger

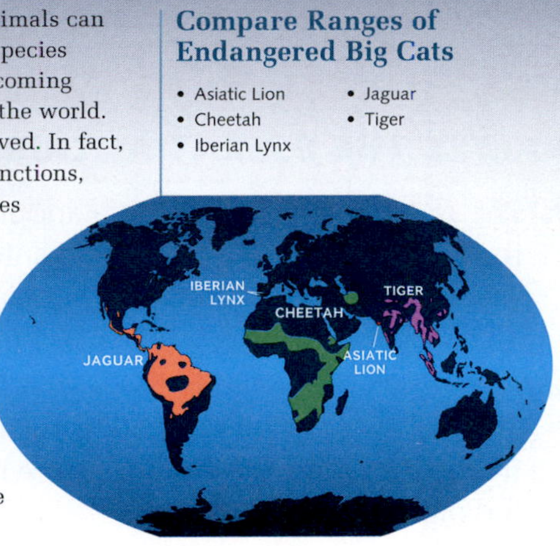

CAUSES OF EXTINCTION

Mining, logging, and clearing of forests for grazing cattle and growing crops all greatly change the natural landscape, or habitat, on which most species depend. Other development, such as the building of dams, highways, and housing, can increasingly divide animal populations into smaller, less diverse pockets. Some groups, like the big cats shown at right, face additional threats from hunters fearful of the animals' ability to harm people and livestock.

Climate change also can seriously stress a species' population and push it to extinction. For example, climate change can alter the amount of rain that falls, which affects plant growth and changes the food available in the habitat. When that happens quickly, species have difficulty adapting.

CONSERVATION

Species and their ecosystems contribute much to the health and well being of humans. A diversity of plants and animals, and the habitats in which they are found, provide fertile soils, medicines, clean air and water, fibers, building materials, and food.

The International Union for Conservation of Animals (IUCN) is one organization trying to help countries around the world find solutions for balancing human needs and environmental challenges. To help target the habitats and animals in need of support, the IUCN maintains a database called the "Red List of Endangered Species." The list identifies animal species as near threatened, vulnerable, endangered, and critically endangered, as shown on the diagram at right.

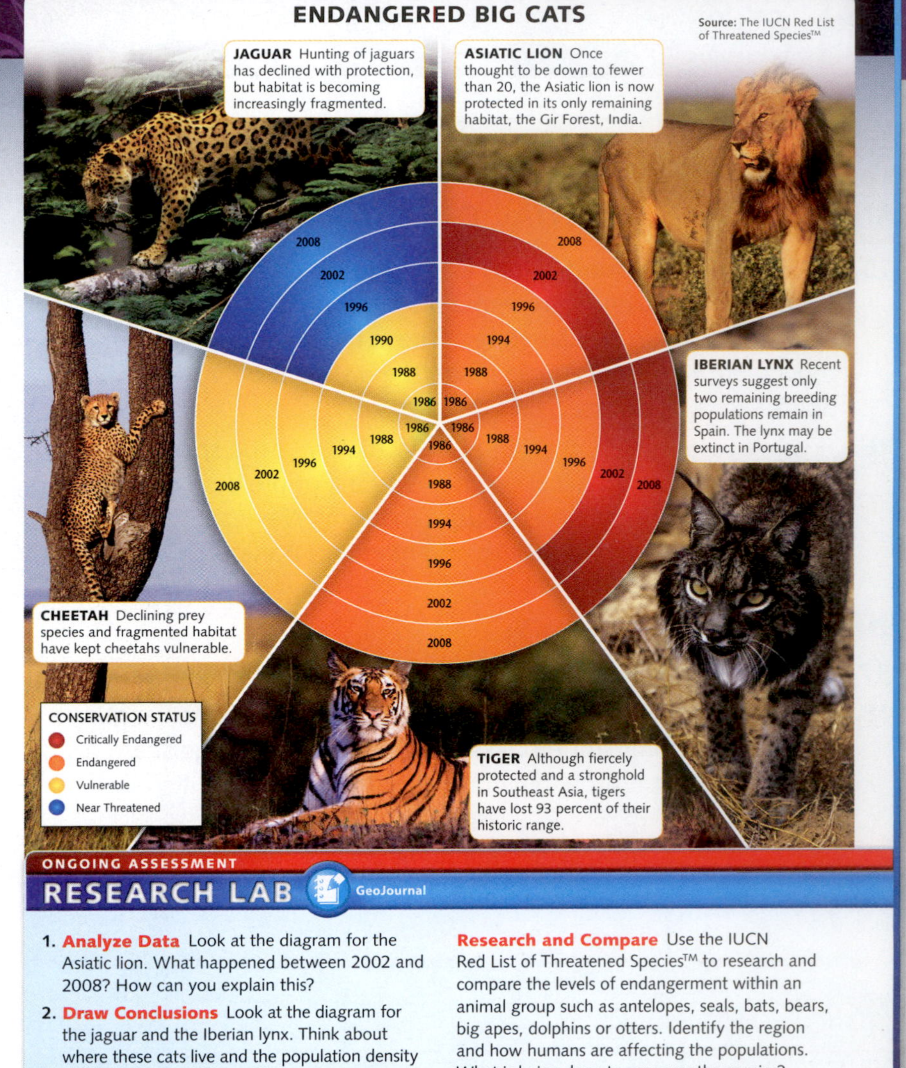

ONGOING ASSESSMENT

RESEARCH LAB

GeoJournal

1. **Analyze Data** Look at the diagram for the Asiatic lion. What happened between 2002 and 2008? How can you explain this?
2. **Draw Conclusions** Look at the diagram for the jaguar and the Iberian lynx. Think about where these cats live and the population density in those ranges. Why might this help explain why their levels of endangerment have changed?

Research and Compare Use the IUCN Red List of Threatened Species™ to research and compare the levels of endangerment within an animal group such as antelopes, seals, bats, bears, big apes, dolphins or otters. Identify the region and how humans are affecting the populations. What is being done to conserve the species? Create a table to describe your findings.

PLAN

OBJECTIVE Compare graphs depicting the population of endangered cats in different countries.

CRITICAL THINKING SKILLS

- Analyze Data
- Draw Conclusions
- Compare
- Analyze Cause and Effect
- Identify Problems and Solutions
- Interpret Graphs

PRINT RESOURCES

Teacher's Edition Resource Bank

Review: GeoActivity Investigate Endangered Species

Maps and Graphs
Graph: Endangered Big Cats

Connect to NG
Research Links

BACKGROUND FOR THE TEACHER

The Iberian lynx makes its home in Spain and Portugal. This 25-pound cat is among the most endangered of species with 225 animals counted in the wild in 2010. Hunting and habitat loss are key reasons for the decline in lynx numbers. Another problem is the cats' nearly exclusive diet of rabbits, which have also seen a severe reduction in numbers due to hunting and disease. The Spanish conservation group Life Lynx is dedicated to protecting the existing Iberian lynxes and increasing their numbers.

Asiatic lions, native to India, have also been reduced in number to a few hundred individuals. These cats were first protected in 1907 by a local ruler who set aside the land for them that later became the Gir National Park. At the time, it was believed that only 13 lions lived in the park. Because of this extreme reduction in numbers, the Asiatic lions alive today have very little diversity in their DNA, which makes them especially vulnerable to diseases.

INTRODUCE & ENGAGE

Identify Causes Explain that scientists have identified five mass extinctions during Earth's history, the most recent being the extinction of the dinosaurs over 65 million years ago. Have pairs brainstorm causes of extinction then and now. As pairs share their lists, have students identify which causes are specific to today. 0:10 minutes

TEACH

Maps and Graphs

Guided Discussion

1. **Analyze Cause and Effect** What effect can building roads and homes have on ecological diversity? Why? *(Building human infrastructure can reduce ecological diversity. Roads, homes, and other structures reduce habitat, crowding out other species. If species become extinct, there is less diversity in the environment.)*
2. **Identify Problems and Solutions** Choose one of the problems threatening the existence of species on Earth. What is a possible solution to the problem? *(Possible response: human development—build fewer new structures and use existing ones; build structures in places where they won't crowd out animal habitat)*

Interpret Graphs Have students look at the Endangered Big Cats graph or project it from **Graphs.** Make sure students can read and understand the graph. **ASK:** What do the color bands tell you about the big cats represented in the graph? *(They tell the endangered status of the species during the years marked in each band.)* What does a change in color tell you about a species? *(that its numbers have gone up or down and its status has changed)* Which cats have had a stable population over the years represented in the graph? *(the tiger and the cheetah)*

Have students use the information in the graph to make a generalization about big cat species. *(Possible response: It's not clear that any progress has been made in protecting the big cat species. Some have seen improvement, some are at greater risk, and some have remained steady.)* 0:15 minutes

DIFFERENTIATE

Research Links

English Language Learners Use Cognates and Context Clues Several of the key concepts in this lesson have Spanish cognates: *extinct–extinto; species–especie; scientist–científico; climate–clima; population–población; habitat–hábitat; adapt–adaptarse*. Except for the cheetah *(guepardo)*, the cat names should be understandable to Spanish speakers as well *(león asiático, león ibérico, tigre, jaguar)*. Have students work in pairs to read the lesson, using these cognates as well as context clues to understand the words. Students may look up words to confirm their guesses as well as words they cannot guess from cognates and context clues.

Pre-AP Conduct Internet Research Have students use the **Research Links** to conduct more in-depth research on one of the species from the IUCN Red List, including the species' habitat and characteristics, the threats it faces, and steps that are being taken to protect it. Students should then create a step-by-step plan for preserving the species. The plan can be presented as a report or using presentation software.

ACTIVE OPTIONS

On Your Feet

Team Word Web Divide the class into groups of four or five and seat each group at a separate table. Give each group a large sheet of butcher paper and give a different-colored marker to each student in each group. Tell half the groups to create a Team Word Web with *extinction* at the center and the other half to create a Team Word Web with *conservation* at the center. Each student adds to the Word Web section in front of him or her. Every minute or so, signal students to rotate the Word Web, allowing each student to work on a different section. When the Word Webs are done, display them in the classroom and give students time to view and compare them. 0:25 minutes

Performance Assessment

Talk-Show Panel Tell students they will be participating in talk-show–style interviews with a host and a panel of experts. Divide the class into four groups and assign each group a role. One group will be the talk-show hosts/interviewers; the other three groups will be the experts. Assign expert groups a topic: extinction, conservation, or big cats. Give the experts time to study their topics and the hosts time to prepare four or five questions. Then place students in groups of four, including one host and one expert on each of the three topics. Have the groups conduct talk-show interviews using their prepared information. You may have students take turns doing their interviews in front of the class, or you may have them do the interviews simultaneously in separate parts of the room as you circulate among the groups.

ONGOING ASSESSMENT

RESEARCH LAB

GeoJournal

ANSWERS

1. The Asiatic lion's threat level decreased. This is probably because of increased protection that occurred within its habitat.
2. The Iberian lynx lives in the more densely populated Iberian peninsula, which includes Spain and Portugal. The jaguar's range includes large areas of rain forest and is largely uninhabited.

Research and Compare Students' research should focus on one animal group and draw clear comparisons between how the species is affected by human activities and what is being done to preserve it.

REVIEW & EXTEND UNIT 11

Active Options

TECHTREK
myNGconnect.com For research links and photos of fruit
Connect to NG | Digital Library | Magazine Maker

ACTIVITY 1

Goal: Learn about unusual and healthy food.

Make a Recommendation

Southeast Asia is home to a rich variety of plants that includes a wide assortment of fruits. Several of the more exotic fruits are listed below. With a partner, use the list below and the research links at Connect to NG to research fruit from the region. Recommend three fruits that your classmates might want to try. Show what the fruits look like and give reasons for recommending those three.

- ciku
- dragon fruit
- durian
- jackfruit
- langsat
- longan
- mangosteen
- rambutan
- salak
- sapodilla
- soursap
- star apple

dragon fruit

ACTIVITY 2

Goal: Research a Southeast Asian city.

Promote a City

Research one of the capital cities below to receive an award for Best Tourist Site of Southeast Asia. Use the **Magazine Maker** to present and describe the site that causes the city to earn the award. Include things to do and places to visit that would delight tourists.

- Bangkok, Thailand
- Kuala Lumpur, Malaysia
- Phnom Penh, Cambodia
- Singapore, Singapore

ACTIVITY 3

Goal: Extend knowledge of Southeast Asian wildlife through research and drama.

Set the Stage

Some of the world's oldest rain forests are in Southeast Asia. With a group, compose a one-act play set deep in the rain forest. Focus the drama on the region's rain forest issues, and include in your play one of the animals found there, such as the Komodo dragon, flying snake, or orangutan. Perform the play for the class.

640 UNIT 11

ASSESS

Use the rubrics to assess each student's participation and performance.

Project Rubric: Activity 1

SCORE	Planning / Preparation	Content / Presentation	Participation / Collaboration
3 GREAT	• Uses two or more reliable sources to research information. • Researches at least five fruits before choosing three to recommend.	• Recommends three fruits; gives several clear and persuasive reasons for each one. • Includes at least one appealing, accurately identified image of each fruit.	• Negotiates roles and responsibilities with the partner; carries out the responsibilities of the chosen roles.
2 GOOD	• Uses one or more reliable sources to research information. • Researches at least four fruits before choosing three to recommend.	• Recommends three fruits; gives at least one reasonable reason for each one. • Includes one accurately identified image of each fruit.	• Talks about roles and responsibilities with the partner, but the decisions may be unclear or student may not fully assume the responsibilities of the chose roles.
1 NEEDS WORK	• Conducts little research or does not use a reliable source. • Researches three or fewer fruits.	• Recommends fewer than three fruits; reasons are lacking or unclear. • Does not have an image for each fruit, or images are not identified.	• Does not discuss roles and responsibilities with the partner; does not assume responsibility for a reasonable share of the work.

Project Rubric: Activity 2

SCORE	Planning / Preparation	Content / Presentation	Participation / Collaboration
3 GREAT	• Uses at least two reliable sources to conduct thorough research on the city. • Focuses research on aspects of the city that would appeal to tourists.	• Takes full advantage of the **Magazine Maker's** features to create an interesting presentation. • The reasons for choosing the city are clear and compelling.	• Views others' presentations attentively and offers constructive feedback.
2 GOOD	• Uses at least one reliable source to conduct adequate research on the city. • Focuses most of the research on aspects of the city that would appeal to tourists.	• Uses the **Magazine Maker's** features to create an interesting presentation. • The reasons for choosing the city are clear.	• Views others' presentations with some attention and offers some helpful feedback.
1 NEEDS WORK	• Conducts little or no research and fails to check reliability of sources. • Research lacks focus and does not pinpoint items with tourist appeal.	• Makes little use of the **Magazine Maker's** features; presentation is dull and hard to follow. • The reasons for choosing the city are unclear.	• Does not pay attention to others' presentations; offers little or no feedback.

Project Rubric: Activity 3

SCORE	Planning / Preparation	Content / Presentation	Participation / Collaboration
3 GREAT	• Conducts thorough, in-depth research on the play's rain forest setting and featured animal.	• Knows the lines perfectly and delivers them with confidence and expression. • The characters are well developed; the plot is compelling and easy to follow.	• Assumes a clear role and related responsibilities. • Actively collaborates with others, offering ideas and acknowledging others' contributions.
2 GOOD	• Conducts sufficient research on the play's rain forest setting and featured animal.	• Knows the lines and delivers them with some expression. • The characters are easy to tell apart; the plot is interesting and easy to follow for the most part.	• Has some difficulty sharing decisions and responsibilities. • Collaborates with others much of the time, sharing some ideas and listening to others' contributions.
1 NEEDS WORK	• Conducts little or no research on the play's rain forest setting; does not choose an animal, or does not research the chosen animal.	• Does not know many of the lines the lines and delivers them mechanically. • The characters are hard to tell apart; the plot is hard to follow.	• Cannot share decisions or responsibilities. • Rarely collaborates with others.

Southeast Asia

Southeast Asia RESOURCE BANK

GEOGRAPHY & HISTORY

SECTION 1 GEOGRAPHY

SECTION 2 HISTORY

TODAY

SECTION 1 CULTURE

SECTION 2 GOVERNMENT & ECONOMICS

FORMAL ASSESSMENT

GEOGRAPHY & HISTORY

TODAY

BLACK-AND-WHITE COPY READY

Name ______ Class ______ Date ______

SECTION 1 GEOGRAPHY

1.1 Physical Geography

Use with Southeast Asia Geography & History, Section 1.1, *in your textbook.*

Reading and Note-Taking Answer Questions

Answer the questions in the Section Map below to help you keep track of information in Section 1.1 about Southeast Asia's physical geography.

What is the title of the section?

Physical Geography

What is the Main Idea?

______

What are the most important details in the "Mainland Countries" section?	What are the most important details in the "Island Countries" section?	What does the map show?
Mainland countries include Myanmar, Thailand, Laos, Vietnam, and Cambodia.		

What does the photograph of the typhoon show?	What does the photograph of the tsunami show?

Compare and Contrast Which country is at greater risk of severe damage from a tsunami—Cambodia or the Philippines? Explain.

______

Name Class Date

SECTION 1 GEOGRAPHY

1.1 Physical Geography

Use with Southeast Asia Geography & History, Section 1.1, *in your textbook.*

Vocabulary Practice

KEY VOCABULARY

- **land bridge** n., a strip of land connecting two land masses
- **landlocked** adj., surrounded by land on all sides
- **tsunami** (su-NAH-mee) n., a giant ocean wave with enormous power
- **typhoon** (ty-FOON) n., a fierce tropical storm with heavy rains and high winds

Comparison Paragraphs Write two paragraphs to compare and contrast the Key Vocabulary words. Write about the relationship between *land bridge* and *landlocked* in the first paragraph, and compare and contrast *tsunami* and *typhoon* in the second paragraph.

Land bridge and landlocked are both used to describe landforms.

BLACK-AND-WHITE COPY READY

Name Class Date

SECTION 1 GEOGRAPHY
GeoActivity

Use with Southeast Asia Geography & History, Section 1.1, *in your textbook.*
Go to Interactive Whiteboard GeoActivities at **myNGconnect.com** *to complete this activity online.*

NATIONAL GEOGRAPHIC School Publishing

1.1 PHYSICAL GEOGRAPHY

Compare Past and Present Land Areas

Read the passage below and study the map, which uses shading to show land areas in Southeast Asia during the last glacial period. Then answer the questions.

Southeast Asia: Past and Present

Glacial Period About 20,000 years ago, much of Earth's water was frozen. Glaciers more than a mile thick covered parts of North America and northern Europe. Because so much water was frozen, sea levels around the world were about 400 feet lower than they are today. Large areas of land were above the water level, which led to the creation of dry land bridges. The first humans in North America crossed from Asia on such a land bridge.

Land Bridges in Southeast Asia Glaciers did not cover Southeast Asia during this period, but the falling sea levels created many land bridges there as well. For example, the Malay Peninsula and several of the islands of Indonesia were once connected by land. Plants, animals, and people could migrate across these land bridges.

Rising Sea Levels As the last glacial period ended about 18,000 years ago, the melting of ice resulted in slowly rising sea levels. About 6,000 years ago, the levels began to stop at present-day levels. Recently, however, sea levels have been rising even more quickly.

1. **Compare Maps** Compare the map of Southeast Asia in Section 1.1 of your textbook to the map shown at right. Which islands of modern Indonesia were linked with mainland Asia during the last glacial period? What major islands were not linked? What might explain this situation?

____

____

SOUTHEAST ASIA: PAST AND PRESENT

100°E 110°E 120°E 130°E 140°E 20°N 10°N 0° 10°S Equator
PHILIPPINE ISLANDS
PACIFIC OCEAN
INDIAN OCEAN
Malay Peninsula
Sumatra
Borneo
Sulawesi (Celebes)
Java
Bali
New Guinea
N S E W

Land areas during the last glacial period about 18,000 years ago
Current extent of land areas
0 300 600 Miles
0 300 600 Kilometers

2. **Make Inferences** What changes could have taken place in this region while the land areas were connected?

____

____

3. **Make Predictions** Modern sea levels are rising. If they continue to rise quickly, what might happen to island countries?

____

____

BLACK-AND-WHITE COPY READY

Name Class Date

SECTION 1 GEOGRAPHY

1.2 Parallel Rivers

Use with Southeast Asia Geography & History, Section 1.2, *in your textbook.*

Reading and Note-Taking Organize Information

As you read Section 1.2, use the table below to help you organize information about the major rivers of Southeast Asia's mainland. Remember to use both the text and the photographs in the section to help you.

Mekong River	Chao Phraya River	Irrawaddy River
Location Myanmar, Laos, Thailand, and Vietnam		
Major Cities Along the River		
Economic Uses		
Other Notable Features		

Name Class Date

SECTION 1 GEOGRAPHY

1.2 Parallel Rivers

Use with Southeast Asia Geography & History, Section 1.2, *in your textbook.*

Vocabulary Practice

KEY VOCABULARY

- **ecologist** (ih-KAH-luh-jihst) n., a scientist who studies the relationship between organisms and their environments
- **subsistence** (suhb-SIHS-tuhns) **fishing** n., the system of catching just enough fish to live on

Definition Chart Complete the Definition Chart for the Key Vocabulary words.

Word	ecologist	subsistence fishing
Definition	a scientist who studies the relationship between organisms and their environments	
In Your Own Words		
Sentence		

BLACK-AND-WHITE COPY READY

Name Class Date

SECTION 1 GEOGRAPHY
GeoActivity

Use with Southeast Asia Geography & History, Section 1.2, *in your textbook.*

Go to Interactive Whiteboard GeoActivities *at* **myNGconnect.com** *to complete this activity online.*

NATIONAL GEOGRAPHIC School Publishing

1.2 PARALLEL RIVERS

Research an Environmental Issue

The Mekong is the longest river in Southeast Asia. It forms parts of the borders of Laos, Myanmar, and Thailand. It flows through Laos, Cambodia, and Vietnam. Many people who live near the Mekong depend on it for water, fishing, transportation, and other needs. However, the river, the region's people, and the region's wildlife face many environmental challenges. Read about a threatened species in the region. Then do some more research on the issue and create a Solution Tree.

An Endangered Giant The Mekong giant catfish faces possible extinction. This freshwater fish can grow to be 10 feet long and weigh 600 pounds. It migrates along the river, traveling to the same place to spawn, or lay eggs, each year. Research why this species is in danger. How can the people of the Mekong region balance economic and environmental concerns?

This giant catfish was caught in Thailand as part of a World Wildlife Fund-National Geographic tracking project. The fish was tagged and released back into the water.

1. **Conduct Research** To learn more about the Mekong giant catfish, work with a teacher or librarian to find online or printed sources. You might begin by using the research links at **Connect to NG**.

2. **Identify Problems** From your research, identify two major problems affecting the survival of the Mekong catfish. Write the problems you have identified in the "Problem" section of the Solution Tree.

3. **Identify Solutions** Work with a partner or small group to suggest possible solutions to each problem. Use information from your research or use your own ideas. Write your solutions in the "Solution" section.

Solution Tree

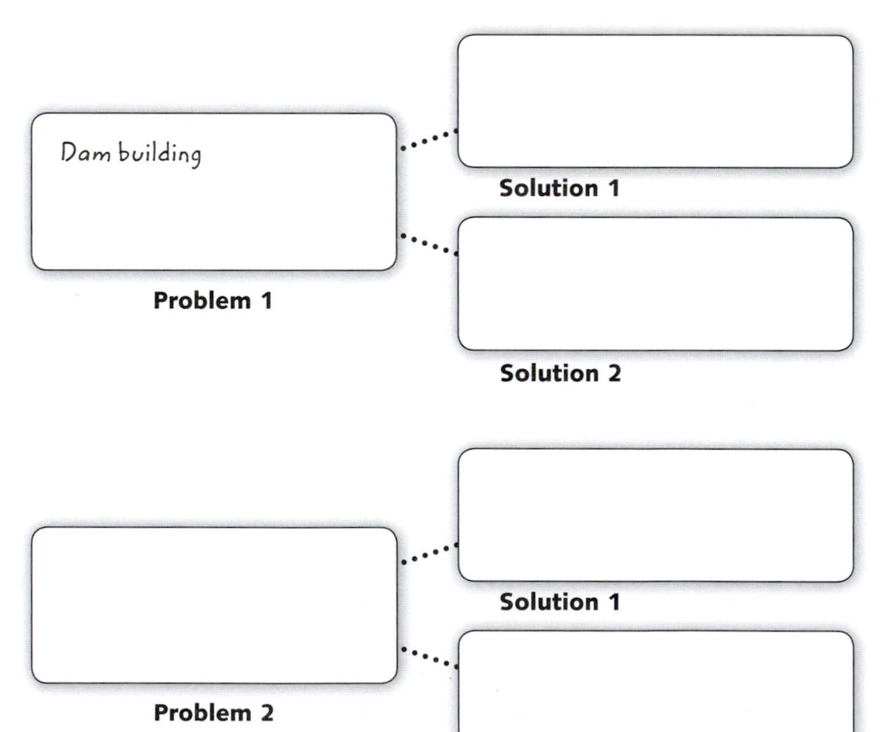

Name Class Date

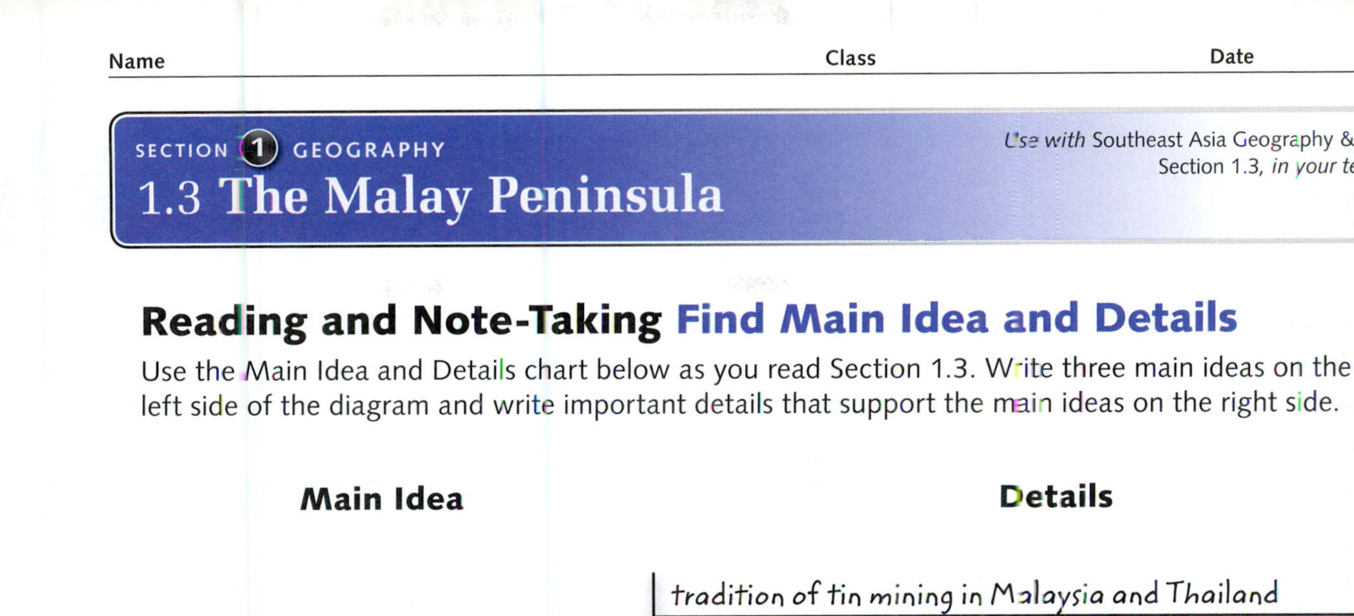

SECTION 1 GEOGRAPHY

1.3 The Malay Peninsula

Use with Southeast Asia Geography & History, Section 1.3, *in your textbook.*

Reading and Note-Taking Find Main Idea and Details

Use the Main Idea and Details chart below as you read Section 1.3. Write three main ideas on the left side of the diagram and write important details that support the main ideas on the right side.

Main Idea	Details
Mineral Resources	tradition of tin mining in Malaysia and Thailand

Analyze How do different industries compete for resources on the Malay Peninsula?

SECTION 1 GEOGRAPHY

1.3 The Malay Peninsula

Use with Southeast Asia Geography & History, Section 1.3, *in your textbook.*

Vocabulary Practice

KEY VOCABULARY

- **bauxite** (BAWK-syt) n., the raw material used to make aluminum
- **biodiversity** (by-oh-duh-VUHR-suh-tee) n., the variety of species in an ecosystem

Word Map Complete a Word Map for the Key Vocabulary words *bauxite* and *biodiversity*. Write the definition for each. To describe what it is like, provide one or two related words or synonyms. Then explain one way the word helps us understand the physical geography of the Malay Peninsula.

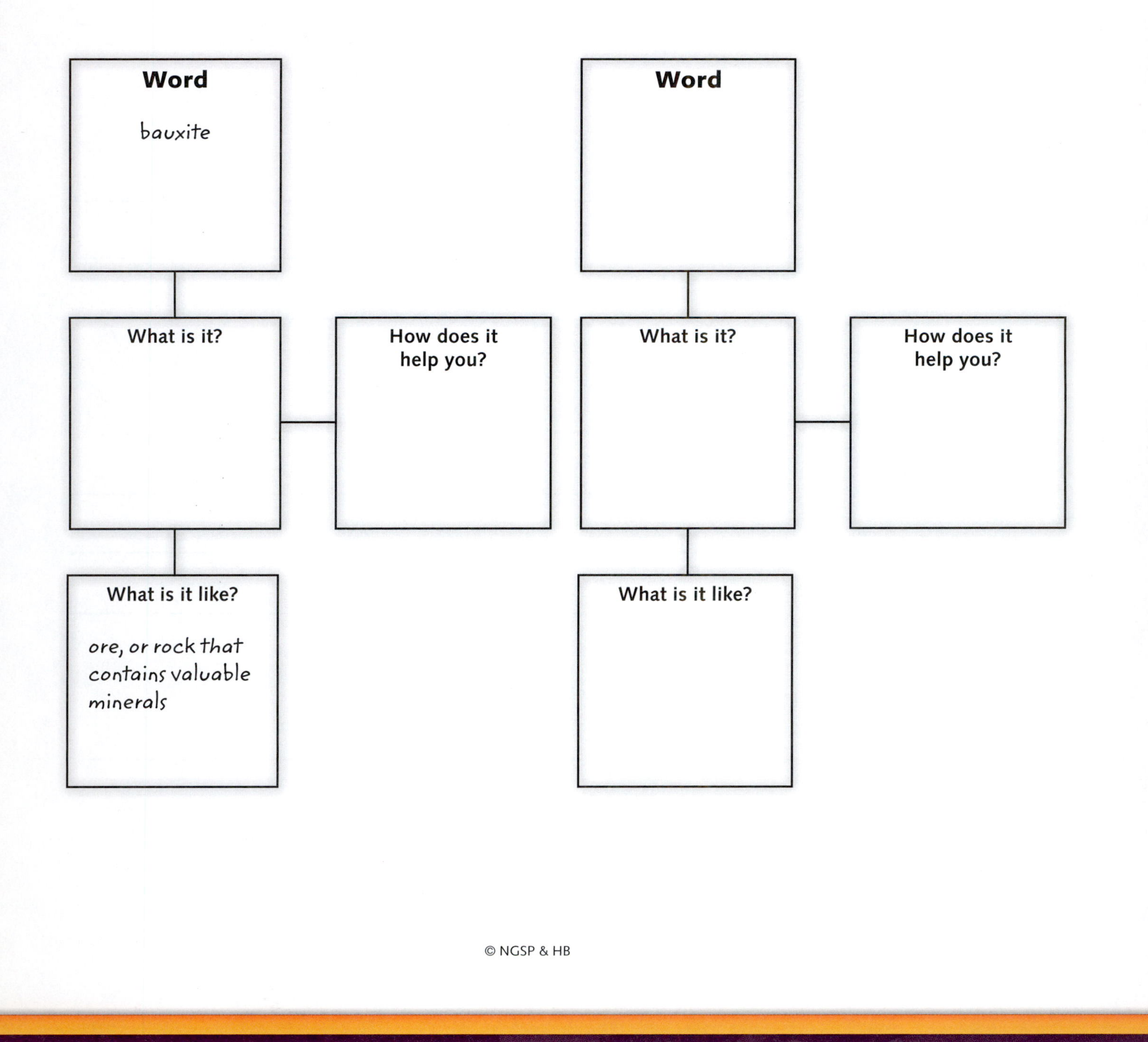

Name ____________ Class ____________ Date ____________

SECTION 1 GEOGRAPHY

GeoActivity

Use with Southeast Asia Geography & History, Section 1.3, *in your textbook.*

Go to Interactive Whiteboard GeoActivities *at* **myNGconnect.com** *to complete this activity online.*

NATIONAL GEOGRAPHIC School Publishing

1.3 THE MALAY PENINSULA

Graph Global Deforestation Rates

Tropical forests all over the world have been gradually disappearing for many years. People clear forest land to make room for agriculture. They cut down trees and use the wood for fuel or for products such as furniture and paper. People sometimes set forest fires to clear the land for farms and settlements.

Deforestation is an environmental crisis in Malaysia. However, logging and the export of timber are important to the country's economy. Review the chart at right, which gives figures for the rates of deforestation in Malaysia and two other countries between 1990 and 2010. Compare the loss of forest cover in Malaysia with that in the other countries, and answer the questions.

1. **Create Graphs** Use the data in the chart at right to construct a line graph comparing deforestation in three countries. Use a different color or pattern for each line. Label it with the country's name.

Deforestation Rates, 1990–2010

RATE OF DEFORESTATION (PERCENT)

2.2
2.0
1.8
1.6
1.4
1.2
1.0
0.8
0.6
0.4
0.2
0

1990–2000 2000–2005 2005–2010

Percentage Loss of Forest Cover

TIME PERIOD	MALAYSIA	BRAZIL	NICARAGUA
1990–2000	0.36	0.51	1.67
2000–2005	0.66	0.57	1.91
2005–2010	0.42	0.42	2.11

Source: The Food and Agriculture Organization of the United Nations

2. **Analyze Data** In which time period was deforestation most serious in Malaysia? What happened during the following time period?

____

____

3. **Analyze Data** What does the data indicate for the deforestation in Nicaragua as compared with the other two countries?

____

____

4. **Make Inferences** What factors might explain the increase or decrease in deforestation in each country?

____

____

5. **Make Generalizations** Malaysia has strong environmental laws, but they often go unenforced. How can the people and the government help with the deforestation crisis?

____

____

Name Class Date

SECTION 1 GEOGRAPHY

1.4 Island Nations

Use with Southeast Asia Geography & History, Section 1.4, *in your textbook.*

Reading and Note-Taking Outline and Take Notes

As you read Section 1.4, use an outline to take notes about the physical geography of Southeast Asia's island nations. Provide three of the section's major ideas and at least two details to support each of these ideas.

I. Dynamic Geographic Zone

A. Multiple plates collide, creating volcanic mountains.

B.

II.

A.

B.

III.

A.

B.

BLACK-AND-WHITE COPY READY

Name Class Date

SECTION 1 GEOGRAPHY

1.4 Island Nations

Use with Southeast Asia Geography & History, Section 1.4, *in your textbook.*

Vocabulary Practice

KEY VOCABULARY

- **dormant** (DAWR-muhnt) adj., inactive for long periods of time
- **dynamic** (dy-NA-mihk) adj., continuously changing or active

ACADEMIC VOCABULARY

- **enhance** (ihn-HANS) v., to increase or improve the value or quality of something

Use the Academic Vocabulary word *enhance* in a sentence about agriculture in Southeast Asia.

Comparison Chart Complete the Y-Chart to compare the meaning of the Key Vocabulary words *dormant* and *dynamic*. Then write how the two words are related.

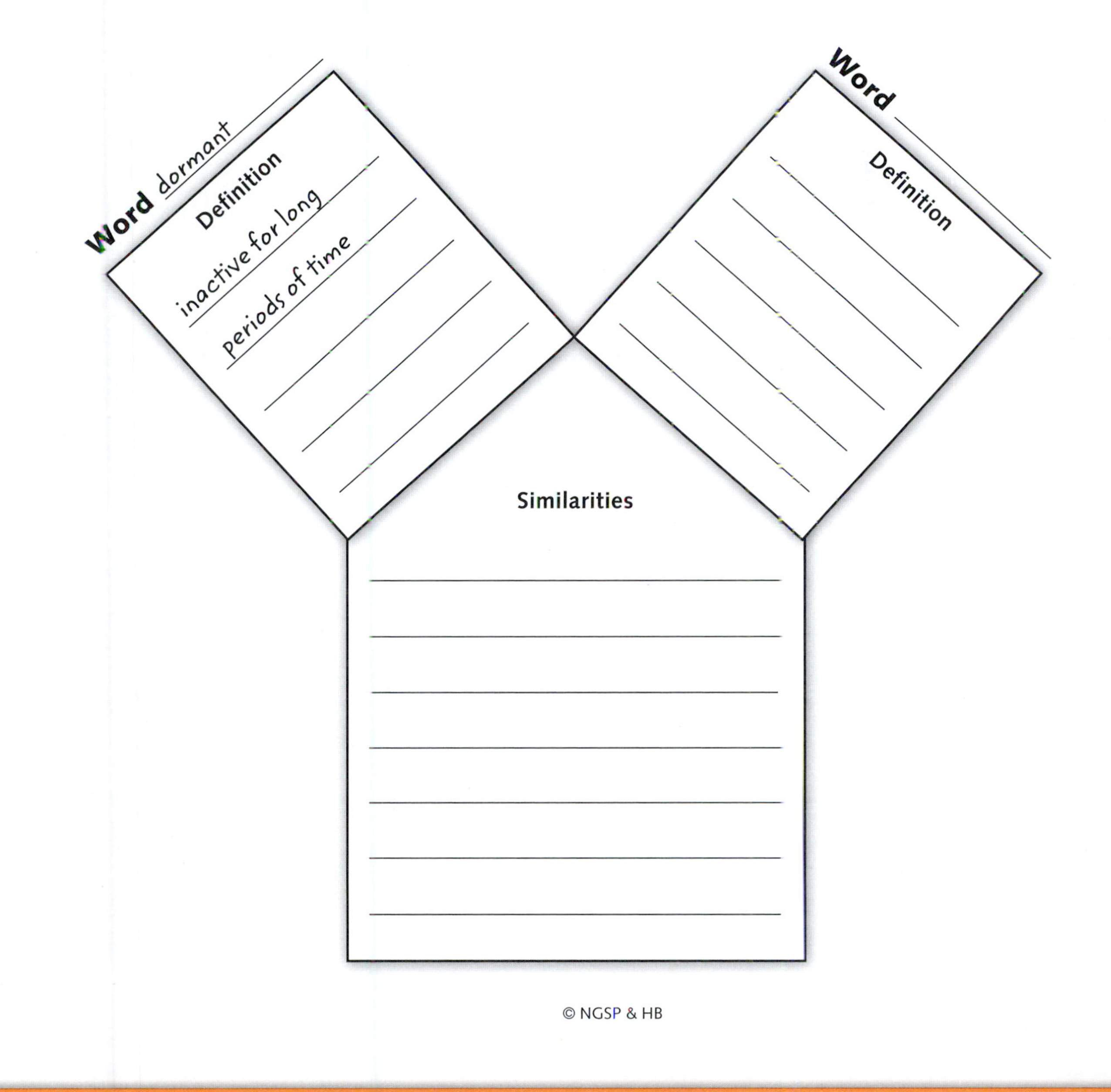

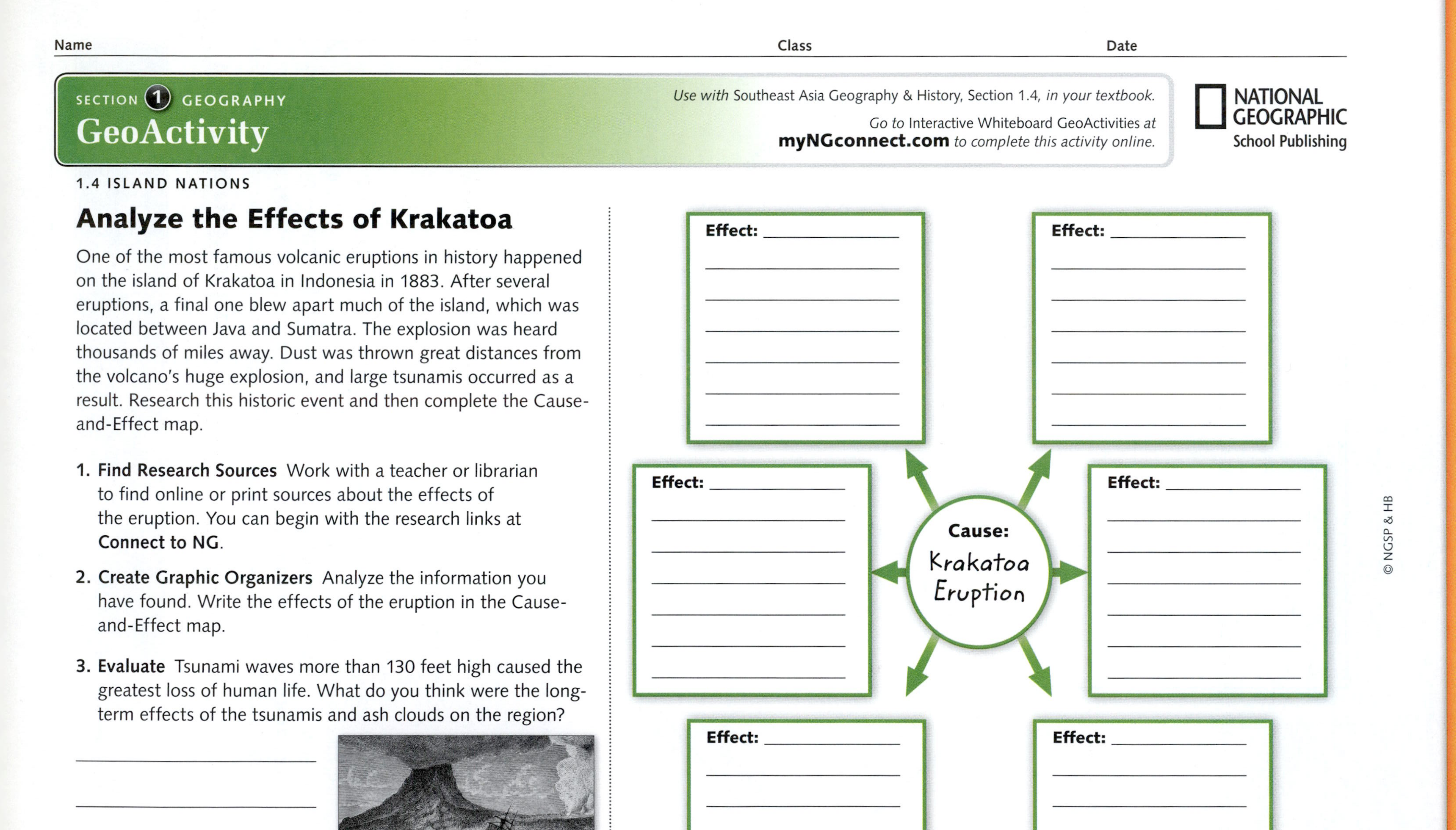

Name ____________________ Class ____________ Date ____________

SECTION 1 GEOGRAPHY

GeoActivity

Use with Southeast Asia Geography & History, Section 1.4, *in your textbook.*

Go to Interactive Whiteboard GeoActivities *at* **myNGconnect.com** *to complete this activity online.*

NATIONAL GEOGRAPHIC School Publishing

1.4 ISLAND NATIONS

Analyze the Effects of Krakatoa

One of the most famous volcanic eruptions in history happened on the island of Krakatoa in Indonesia in 1883. After several eruptions, a final one blew apart much of the island, which was located between Java and Sumatra. The explosion was heard thousands of miles away. Dust was thrown great distances from the volcano's huge explosion, and large tsunamis occurred as a result. Research this historic event and then complete the Cause-and-Effect map.

1. **Find Research Sources** Work with a teacher or librarian to find online or print sources about the effects of the eruption. You can begin with the research links at **Connect to NG**.
2. **Create Graphic Organizers** Analyze the information you have found. Write the effects of the eruption in the Cause-and-Effect map.
3. **Evaluate** Tsunami waves more than 130 feet high caused the greatest loss of human life. What do you think were the long-term effects of the tsunamis and ash clouds on the region?

A 19th-century engraving of the Krakatoa explosion

Cause: Krakatoa Eruption

Effect: ____________

Effect: ____________

Effect: ____________

Effect: ____________

Effect: ____________

Effect: ____________

Name Class Date

SECTION 1 GEOGRAPHY

1.5 Discovering New Species

Use with Southeast Asia Geography & History, Section 1.5, *in your textbook.*

Reading and Note-Taking Summarize Information

As you read Section 1.5, use the Idea Web below to help you take notes about the discovery of new species in the Foja Mountains. Write the section's main idea in the central square and summarize the details in the outer circles.

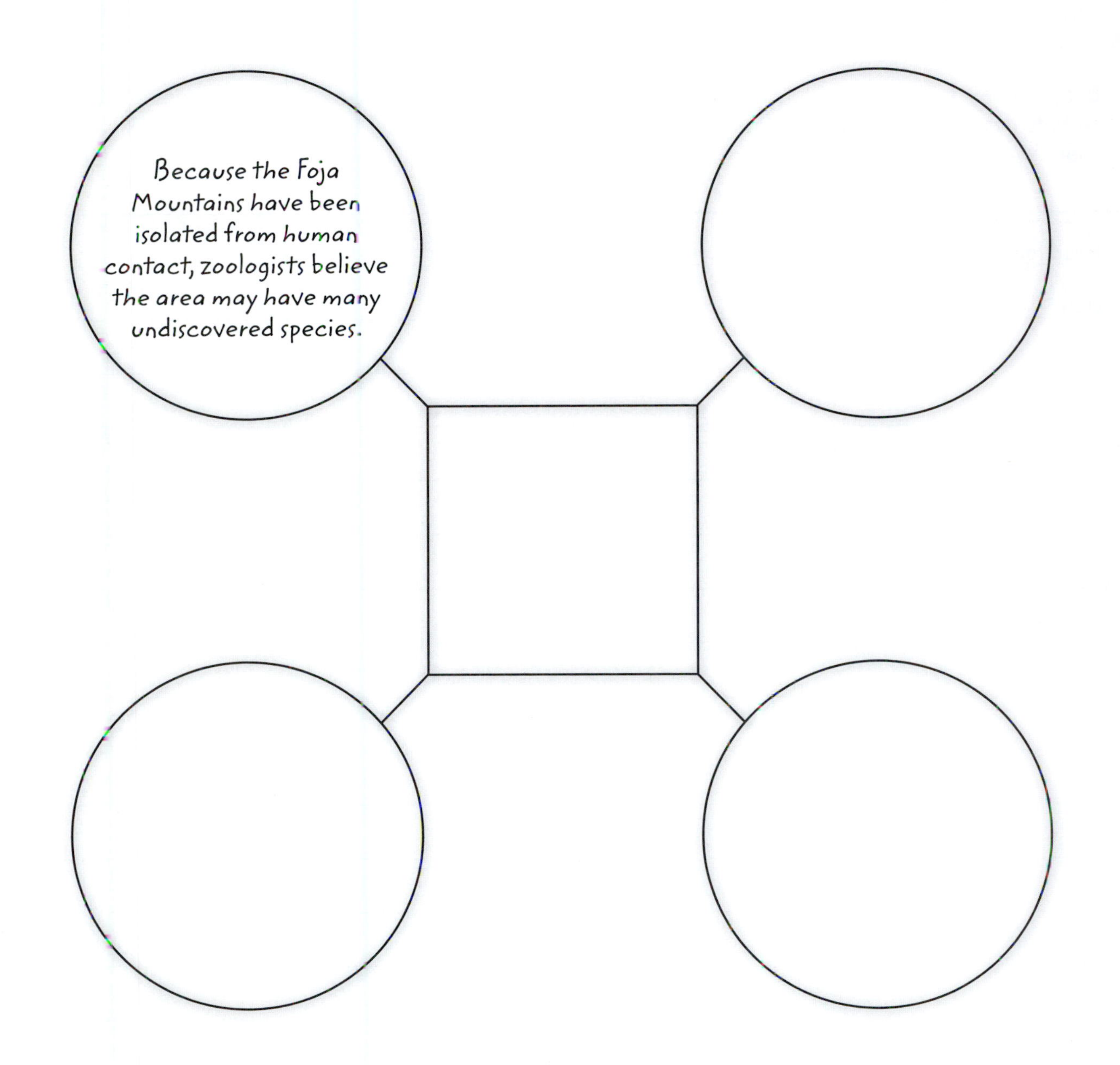

Make Inferences Do you think the Foja Mountains are easily accessible? Explain your answer.

Name Class Date

SECTION 1 GEOGRAPHY

1.5 Discovering New Species

Use with Southeast Asia Geography & History, Section 1.5, *in your textbook.*

Vocabulary Practice

KEY VOCABULARY

- **wallaby** (WAH-luh-bee) n., a smaller relative of the kangaroo
- **zoologist** (zoh-AH-luh-jihst) n., a scientist who studies animals

Blog Entry Imagine you are joining Kristofer Helgen on an expedition to the Foja Mountains. Write a blog entry to describe the day's events. Be sure to provide a headline and date. Use the Key Vocabulary words in your blog entry.

Headline: ______________________________

Date: ______________________________

BLACK-AND-WHITE COPY READY

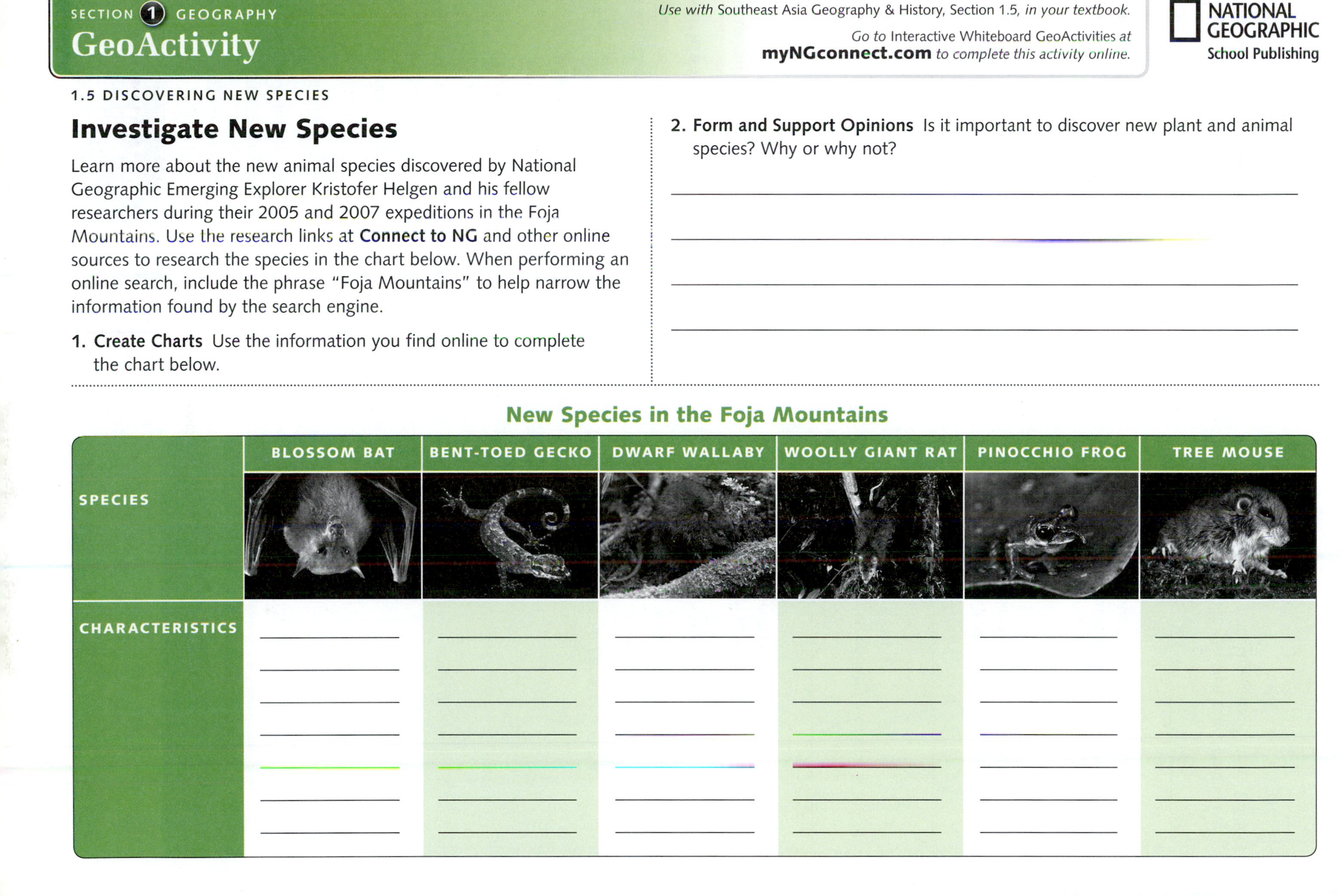

Name Class Date

SECTION 1 GEOGRAPHY
GeoActivity

Use with Southeast Asia Geography & History, Section 1.5, *in your textbook.*

Go to Interactive Whiteboard GeoActivities *at* **myNGconnect.com** *to complete this activity online.*

NATIONAL GEOGRAPHIC School Publishing

1.5 DISCOVERING NEW SPECIES

Investigate New Species

Learn more about the new animal species discovered by National Geographic Emerging Explorer Kristofer Helgen and his fellow researchers during their 2005 and 2007 expeditions in the Foja Mountains. Use the research links at **Connect to NG** and other online sources to research the species in the chart below. When performing an online search, include the phrase "Foja Mountains" to help narrow the information found by the search engine.

1. **Create Charts** Use the information you find online to complete the chart below.

2. **Form and Support Opinions** Is it important to discover new plant and animal species? Why or why not?

New Species in the Foja Mountains

SPECIES	BLOSSOM BAT	BENT-TOED GECKO	DWARF WALLABY	WOOLLY GIANT RAT	PINOCCHIO FROG	TREE MOUSE
CHARACTERISTICS						

Name Class Date

SECTION 1 GEOGRAPHY

1.1–1.5 Review and Assessment

Use with Southeast Asia Geography & History, Sections 1.1–1.5, *in your textbook.*

Follow the instructions below to review what you have learned in this section.

Vocabulary Next to each vocabulary word, write the letter of the correct definition.

1.____ landlocked
2.____ typhoon
3.____ tsunami
4.____ ecologist
5.____ bauxite
6.____ biodiversity
7.____ dynamic
8.____ dormant
9.____ zoologist
10.____ wallaby

A. a fierce tropical storm with heavy rains and high winds
B. continuously changing
C. a smaller relative of the kangaroo
D. inactive for long periods of time
E. surrounded by land on all sides
F. the variety of species in an ecosystem
G. a giant ocean wave with enormous power
H. a scientist who studies animals
I. the raw material used to make aluminum
J. a scientist who studies the relationship between organisms and their environments

Main Ideas Use what you've learned about Southeast Asia's geography to answer these questions.

11. **Categorize** Are Myanmar and Laos island nations or mainland nations?

____

12. **Make Generalizations** What type of climate is found across most of Southeast Asia?

____

13. **Identify** What type of natural disaster struck Sumatra in 2004?

____

14. **Explain** Why is the Chao Phraya River so important to Thailand?

____

15. **Compare and Contrast** Which river is the longest—the Mekong, the Chao Phraya, or the Irrawaddy?

____

16. **Describe Geographic Information** How has the Irrawaddy delta been changing?

____

17. **Identify** Part of what country is on the island of Borneo?

____

18. **Compare and Contrast** In what way are settlement patterns in the Philippines and Indonesia similar?

____

19. **Describe Geographic Information** What type of ecosystem is found in the Foja Mountains?

____

20. **Evaluate** Why is the research conducted in the Foja Mountains particularly important for training zoologists?

____

BLACK-AND-WHITE COPY READY

BLACK-AND-WHITE COPY READY

Focus Skill: Analyze Cause and Effect

Answer the questions about the conditions that have divided Southeast Asia into many different parts.

21. Why does the land bridge that once connected Indonesia to the mainland no longer exist?

____

____

22. Why must ports along the Irrawaddy have two areas for docking?

____

____

23. What effects can building dams along Southeast Asia's rivers have?

____

____

24. Why has mining in Malaysia declined since the 1970s?

____

____

25. What is one consequence of excessive deforestation?

____

____

26. Many farmers live near volcanoes. What are the advantages and disadvantages of this location?

____

____

27. How does climate affect agriculture in the island nations?

____

____

28. Why is the lack of human presence in the Foja Mountains important to scientists?

____

____

Synthesize: Answer the Essential Question

What are the geographic conditions that divide Southeast Asia into many different parts? Consider the information you have learned about Southeast Asia and the geographic barriers that are found both between and within the region's countries.

____

____

____

____

____

____

____

____

____

SECTION 1 GEOGRAPHY

1.1–1.5 Standardized Test Practice

Use with Southeast Asia Geography & History, Sections 1.1–1.5, *in your textbook.*

Follow the instructions below to practice test-taking on what you've learned from this section.

Multiple Choice Circle the best answer for each question from the options available.

1. Soil on Thailand's Central Plain is ideal for growing which crop?
   - A wheat
   - B barley
   - C rice
   - D corn

2. Which of the following is caused by undersea earthquakes?
   - A typhoons
   - B monsoons
   - C droughts
   - D tsunamis

3. The Mekong River forms part of the border of which country?
   - A Thailand
   - B Malaysia
   - C Indonesia
   - D the Philippines

4. Which city is located on the Chao Phraya River?
   - A Ho Chi Minh City
   - B Jakarta
   - C Bangkok
   - D Manila

5. How has Malaysian mining changed since the 1970s?
   - A Large gold deposits have been discovered.
   - B Tin and bauxite deposits have been depleted.
   - C Copper has become much more valuable on the international market.
   - D Iron mining operations have been privatized.

6. Why have sections of the rain forest in Malaysia and Thailand been cut down?
   - A to make room for large farms of certain trees
   - B to clear space for new cities
   - C to make transportation across these countries easier
   - D to create space for livestock to graze

7. The collision of tectonic plates forms which of the following?
   - A coral reefs
   - B mineral deposits
   - C river deltas
   - D volcanoes

8. Which of the following is true of Java?
   - A It is the largest and most populous Indonesian island.
   - B It is smaller than other Indonesian islands but has the largest population.
   - C It is the largest Indonesian island but has a small population.
   - D It has a small land area and a small population.

9. Which area has no human population?
   - A Foja Mountains
   - B Irrawaddy River Basin
   - C Malay Peninsula
   - D Mekong Delta

10. The wallaby is related to what larger animal?
    - A elephant
    - B rhinoceros
    - C kangaroo
    - D tiger

Document-Based Question

The following article, **"Tsunami, Volcano Eruption in Indonesia Linked?"** by Richard A. Lovett, appeared on *National Geographic News* on October 27, 2010. Read the passage and answer the questions below.

> *This week's Indonesian tsunami and volcano eruption might be linked, scientists say. . . . Mount Merapi, Indonesia's most active volcano, had been building up steam for several days. But the timing of the main burst so soon after the earthquake raises the question of whether the shaking ground set off the eruption—even though the epicenter of the quake was 800 miles (1,300 kilometers) away from the volcano.*

Constructed Response Read each question carefully and write your answer in the space provided.

11. According to the passage, why do scientists suspect the tsunami and eruption might be linked?

____

12. How do we know that the earthquake was not the sole cause of the volcanic eruption?

____

Extended Response Read each question carefully and write your answer in the space provided.

13. If the eruption of Mount Merapi was related to the Indonesian earthquake occurring 800 miles away, what geological connection might there be between the earthquake's epicenter and the volcano?

____

____

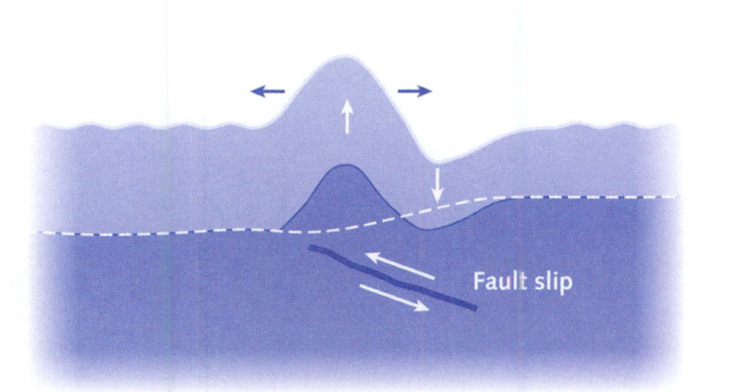

In this diagram an earthquake is occurring under the ocean. When the two sides of the fault slip past each other, it pushes the seafloor up and down, along with the column of water on top of it, causing massive waves that can create tsunamis.

14. The type of movement that can occur along a fault varies. Why would up-and-down movement of two sides of a fault be more likely to produce a tsunami than side-to-side movement along a fault?

____

____

15. How do you think the number of potentially dangerous fault lines in Southeast Asia compares with other parts of the world? Why?

____

____

____

Name Class Date

SECTION 2 HISTORY

2.1 Ancient Valley Kingdoms

Use with Southeast Asia Geography & History, Section 2.1, *in your textbook.*

Reading and Note-Taking Categorize Information

As you read Section 2.1, use the Classification Chart below to help you take notes about Southeast Asia's ancient kingdoms. Dedicate each box to a different kingdom or empire and list two or three notable details about each.

Compare and Contrast In what ways are the Khmer Empire and the Srivijaya Empire similar and different?

Name Class Date

SECTION 2 HISTORY

2.1 Ancient Valley Kingdoms

Use with Southeast Asia Geography & History, Section 2.1, *in your textbook.*

Vocabulary Practice

KEY VOCABULARY

- **bas-relief** (BAH-rih-LEEF) n., a sculpture that slightly projects from a flat background
- **complex** (KAHM-plehks) n., a set of interconnected buildings

Travel Brochure Create a travel brochure for visitors to Angkor Wat. Draw a picture of the temple in the box. Use both Key Vocabulary words in a paragraph that describes the site.

Name ____________________ Class ____________ Date ____________

SECTION 2 HISTORY

GeoActivity

Use with Southeast Asia Geography & History, Section 2.1, *in your textbook.*

Go to Interactive Whiteboard GeoActivities *at* **myNGconnect.com** *to complete this activity online.*

NATIONAL GEOGRAPHIC School Publishing

2.1 ANCIENT VALLEY KINGDOMS

Solve a Puzzle About Ancient Kingdoms

Complete the crossword puzzle and see how much you remember about the ancient valley kingdoms of Southeast Asia. Then unscramble letters to find the Mystery Word.

Crossword Puzzle Tips

1. Read the clues and fill in all the answers you know.
2. Read through the clues again and see if the words you filled in give you clues to other words.
3. If an answer has two words, do not put a space between the words. If an answer includes a punctuation mark, do not include the mark.
4. Look at Section 2.1 in your textbook only if you cannot think of all the answers.

Across

1 Indonesian dynasty that flourished from A.D. 780 to 850

2 Civilization that influenced the cultural direction of Southeast Asia

4 Kingdom established in Southeast Asia in A.D. 939

7 Temple dedicated to the Hindu god Vishnu

8 Religion that reached Southeast Asia around 111 B.C.

9 Largest and longest lasting empire in Southeast Asia

Down

1 Early Indonesian empire that controlled the Strait of Malacca

3 Former capital city on the island of Java

5 Buddhist temple on the island of Java

6 Type of sculpture that projects slightly from a flat background

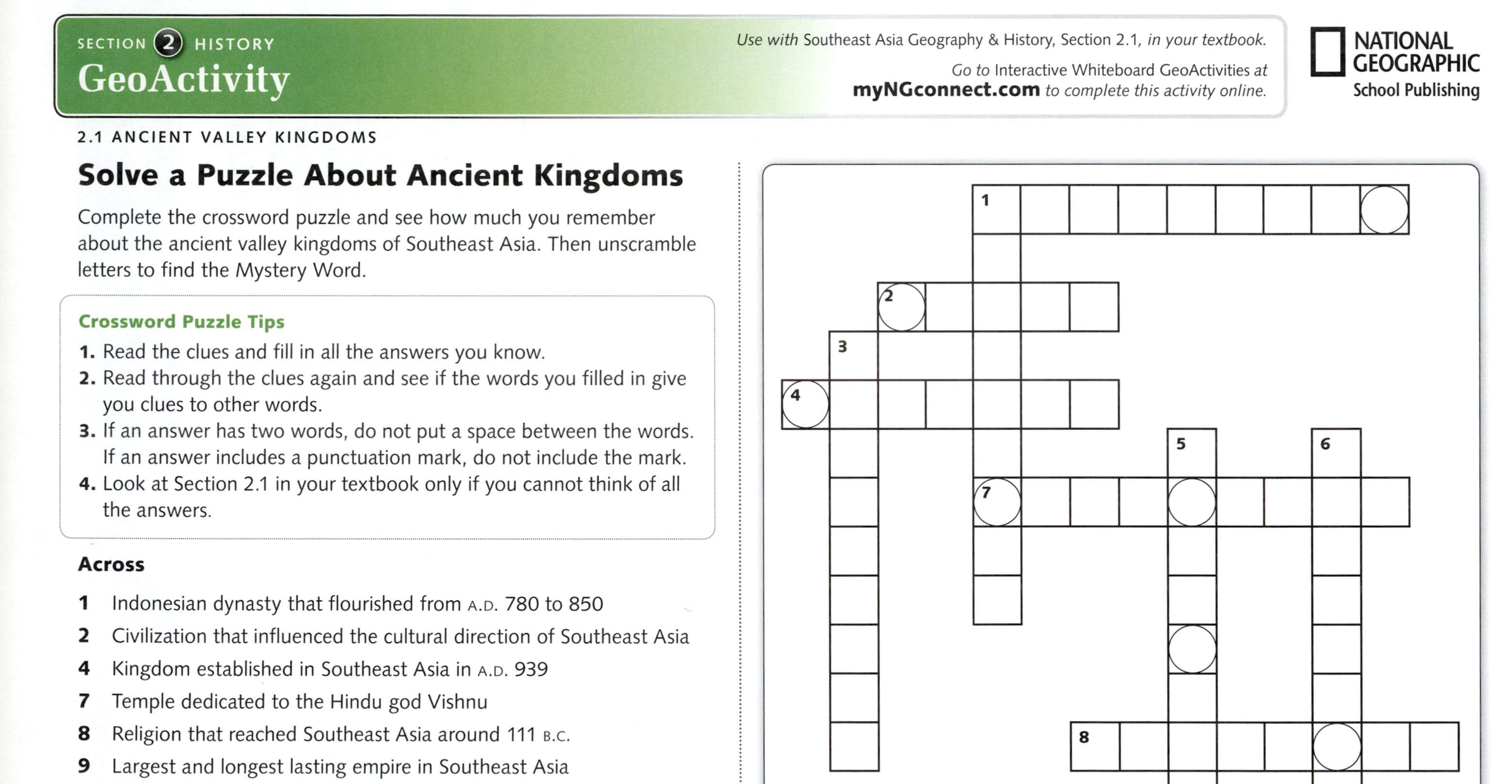

Solve the Mystery Word When you are finished solving the puzzle, look at the circled letters. Rearrange them to spell the name of a country in the region.

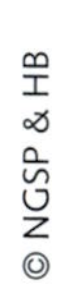

BLACK-AND-WHITE COPY READY

Name Class Date

SECTION 2 HISTORY

2.2 Trade and Colonialism

Use with Southeast Asia Geography & History, Section 2.2, *in your textbook.*

Reading and Note-Taking Draw Conclusions

Use the Idea Diagram below as you read Section 2.2 to help you draw a conclusion about colonialism in Southeast Asia. Write details about each of the colonies. Be sure to provide a conclusion about the effect that colonialism had on both the local populations and on the European powers.

Topic: Colonialism in Southeast Asia

Introduction: ______

Spanish colonies	Dutch colonies	British colonies	French colonies

Details:

Spanish colonies	Dutch colonies	British colonies	French colonies
Spain claimed the Philippines in the 1500s and retained control of the islands until 1898.			

Conclusion: ______

Name Class Date

SECTION 2 HISTORY

2.2 Trade and Colonialism

Use with Southeast Asia Geography & History, Section 2.2, *in your textbook.*

Vocabulary Practice

KEY VOCABULARY

- **colonialism** (kuh-LOH-nee-uh-lih-zuhm) n., one country ruling and developing trade in another country for its own benefit
- **monopoly** (muh-NAH-puh-lee) n., the complete control of a market

Definition Clues Follow the instructions below for the Key Vocabulary word indicated.

Vocabulary Word: colonialism

1. Write the sentence in which the word appears in the section.

2. Write the definition using your own words.

3. Use the word in a sentence of your own.

4. *Colonialism* ends in "-ism." What do you think this suffix means? What is another example of a word ending in "-ism"?

Vocabulary Word: monopoly

1. Write the sentence in which the word appears in the section.

European influence arrived in the 1500s, as merchants hoped to establish a spice trade monopoly, or complete control of the market.

2. Write the definition using your own words.

3. Use the word in a sentence of your own.

4. In a *monopoly*, one group or company controls the entire market. What effect do you think this has on the price of goods?

BLACK-AND-WHITE COPY READY

Name Class Date

SECTION 2 HISTORY
GeoActivity

Use with Southeast Asia Geography & History, Section 2.2, *in your textbook.*
Go to Interactive Whiteboard GeoActivities *at* **myNGconnect.com** *to complete this activity online.*

NATIONAL GEOGRAPHIC School Publishing

2.2 TRADE AND COLONIALISM

Map the Spice Trade

Read the passage below about the spice trade in South Asia and Southeast Asia. Then locate the sources of each spice on the map and answer the questions.

The Spice Trade

Spices such as pepper and cloves were important commodities, or goods, in late medieval Europe. Europeans wanted spices in order to preserve food and improve its taste. Spices grew naturally in parts of Asia and were traded there, but the long journey to bring them to European markets made them expensive luxuries.

Cloves Small, spicy cloves grew naturally only in the Moluccas, a group of islands in northern Indonesia, west of New Guinea. To keep their monopoly on cloves, the Dutch put strict controls on where and how clove-producing evergreen trees could be grown.

Cinnamon Fragrant cinnamon came from the bark of a tropical evergreen that grew naturally in Sri Lanka, in Myanmar, and on the western coast of India. For the Dutch East India Company, cinnamon was the most valuable part of the spice trade.

Nutmeg This spicy seed also grew naturally on tall trees in the Moluccas. Dutch planters tried to keep a monopoly on nutmeg while the French and English tried to get seeds to plant. The outer coating of the seed produces another spice, called *mace*.

Black Pepper The hot, spicy berries of black pepper grew on vines that were native to the western coast of India.

Trade Routes Before Europeans found a sea route around Africa, spices were shipped from the Moluccas to Malacca, and then to Calicut on the southwest coast of India. From there ships headed to the Red Sea.

1. **Create Maps** From the information in the passage, locate the islands or regions that were the source of each spice. Create a symbol for each spice and place it in the appropriate location. Then draw arrows to show possible trade routes to Europe.

SPICE TRADE IN SOUTHEAST ASIA

Cinnamon
Cloves
Nutmeg
Black Pepper

0 500 1,000 Miles
0 500 1,000 Kilometers

2. **Draw Conclusions** The Moluccas became known as the Spice Islands. How is this name appropriate?

3. **Make Predictions** How might European traders have improved their profits from the spice trade besides controlling where spice trees were planted?

BLACK-AND-WHITE COPY READY

Name Class Date

SECTION 2 HISTORY

2.3 Indonesia and the Philippines

Use with Southeast Asia Geography & History, Section 2.3, *in your textbook.*

Reading and Note-Taking Compare and Contrast

As you read Section 2.3, use the Venn Diagram below to keep track of the similarities and differences between Indonesia and the Philippines.

Indonesia

possible home to earliest humans, as suggested by fossils

Similarities

controlled by the Spanish beginning in the 1500s

The Philippines

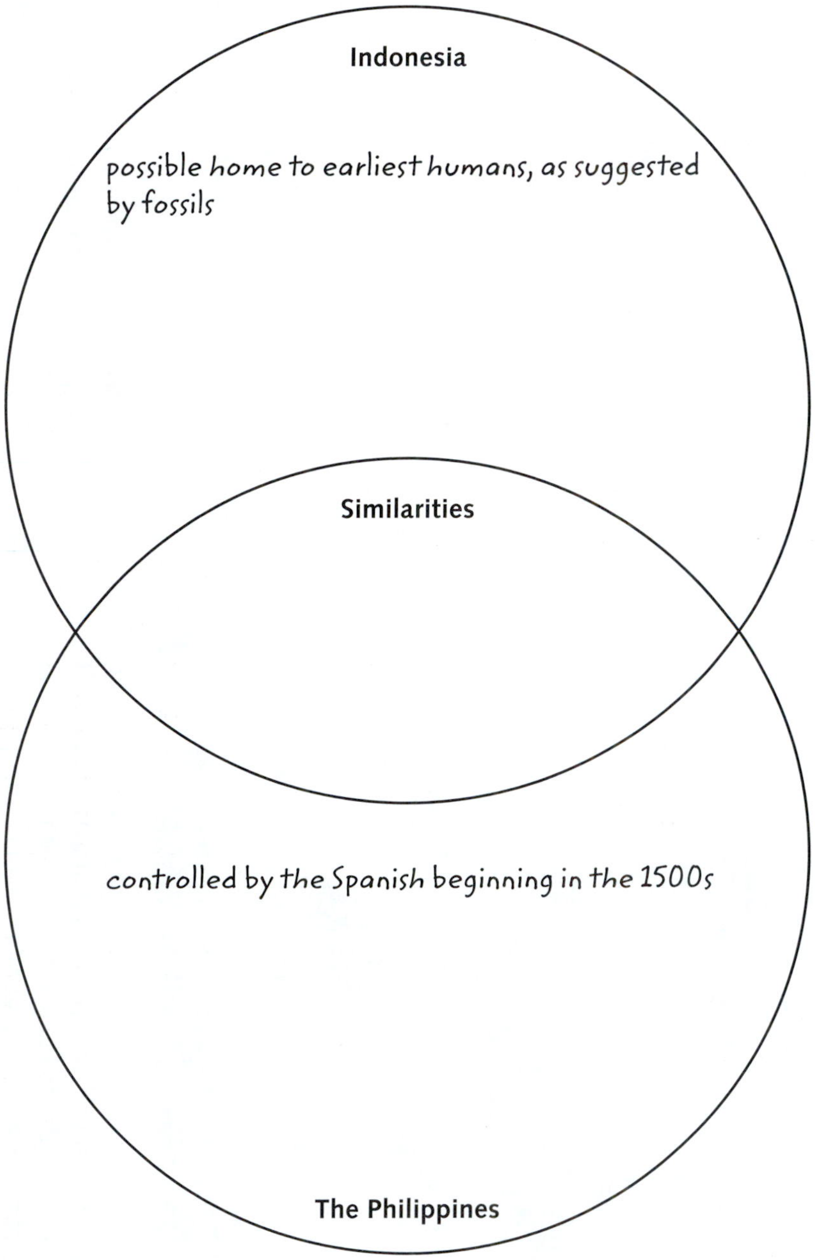

How did Spain's control over the Philippines in the 1800s compare with the Dutch control over Indonesia?

____

____

____

BLACK-AND-WHITE COPY READY

Name Class Date

SECTION 2 HISTORY

2.3 Indonesia and the Philippines

Use with Southeast Asia Geography & History, Section 2.3, *in your textbook.*

Vocabulary Practice

KEY VOCABULARY

- **commerce** (KAH-muhrs) n., the business of trading goods and services
- **fossil** (FAH-suhl) n., the preserved remains or impression of a life form from an earlier time

Words in Context Follow the directions below for using the Key Vocabulary in writing.

1. Explain how the concept of *commerce* relates to the Chinese population in the Philippines.

2. Write a sentence using the word *commerce.*

3. What word or words do you know that are related to the word *commerce*?

4. What types of creatures do you most often associate with *fossils*?

5. How are *fossils* different from artifacts?

6. How do fossils help expand our knowledge of human history?

Name ______________________ Class ______________ Date ______________

SECTION 2 HISTORY
GeoActivity

Use with Southeast Asia Geography & History, Section 2.3, *in your textbook.*
Go to Interactive Whiteboard GeoActivities *at* **myNGconnect.com** *to complete this activity online.*

NATIONAL GEOGRAPHIC School Publishing

2.3 INDONESIA AND THE PHILIPPINES

Analyze Achievements of Emilio Aguinaldo

Emilio Aguinaldo was an important figure in the fight for Filipino independence. Read the following short biography of Aguinaldo and then answer the questions.

A Revolutionary Leader

Emilio Aguinaldo was born in the Philippines in 1869. He was educated in Manila and then became mayor of his hometown. At that time, he also joined the Katipunan, a revolutionary group that fought against Spanish occupation. After two years of revolts against the Spanish, Aguinaldo was sent into exile in Hong Kong in 1898.

While he was in exile, war broke out between the United States and Spain. Aguinaldo returned to the Philippines to fight on the side of the United States. He believed the United States would help the Philippines gain independence. Upon his return, Aguinaldo declared the Philippines' independence from Spain. He was elected president of the new republic.

Neither Spain nor the United States recognized the independence, and, in the treaty that ended the Spanish-American war, Spain ceded, or gave, the Philippines to the United States. Aguinaldo organized a revolution in response. It lasted three years until Aguinaldo was captured by U.S. forces.

The United States made the Philippines a protectorate in 1935. Aguinaldo ran for president but was defeated and retired to private life. His dream of Filipino independence finally came true on July 4, 1946. Although he never held an elected position again, Aguinaldo served his country by working to promote nationalism and democracy until his death in 1964.

Aguinaldo in 1949

1. **Create Charts** Think about the obstacles and achievements Aguinaldo experienced during his struggle for independence. Then fill in the chart below.

OBSTACLE	ACHIEVEMENT

2. **Synthesize** Why did the United States have an interest in keeping the Philippines as its colony? Think about what you learned in Section 2.3 of your textbook as you formulate your response.

____

____

____

____

____

____

BLACK-AND-WHITE COPY READY

Name _______________ Class _______________ Date _______________

SECTION 2 HISTORY

2.4 The Vietnam War

Use with Southeast Asia Geography & History, Section 2.4, *in your textbook.*

Reading and Note-Taking Analyze Primary Sources

Use the chart below to take notes as you read the primary sources in Section 2.4 and evaluate the points of view they represent. Use your knowledge of Southeast Asia's history to help you understand the writers' points of view. Then answer the questions about Document 3.

	Document 1	Document 2
Who wrote this document?		
What is the main idea of this document?	*The United States is determined not to leave Vietnam. If the U.S. did retreat, the forces of communism would have to be fought in other countries.*	
What is the writer's point of view?		
Why do you think the writer feels this way?		

Document 3 What do you see in the photo? How does the photo illustrate the challenges of traveling over Vietnam's terrain for U.S. soldiers?

____

____

Name Class Date

SECTION 2 HISTORY

2.4 The Vietnam War

Use with Southeast Asia Geography & History, Section 2.4, *in your textbook.*

Vocabulary Practice

KEY VOCABULARY

- **launch** v., to start something or set it in motion
- **resistance** (rih-ZIHS-tuhns) n., opposition, or a group that rebels or opposes

ACADEMIC VOCABULARY

- **transform** v., to change or convert something

Write a sentence about the war in Vietnam using the Academic Vocabulary word *transform.*

____

____

Word Squares Complete a Word Square for the Key Vocabulary words *launch* and *resistance*. Write the definition and characteristics or features that help you understand the concept. Give a real-world example and an example of the word's antonym, or opposite.

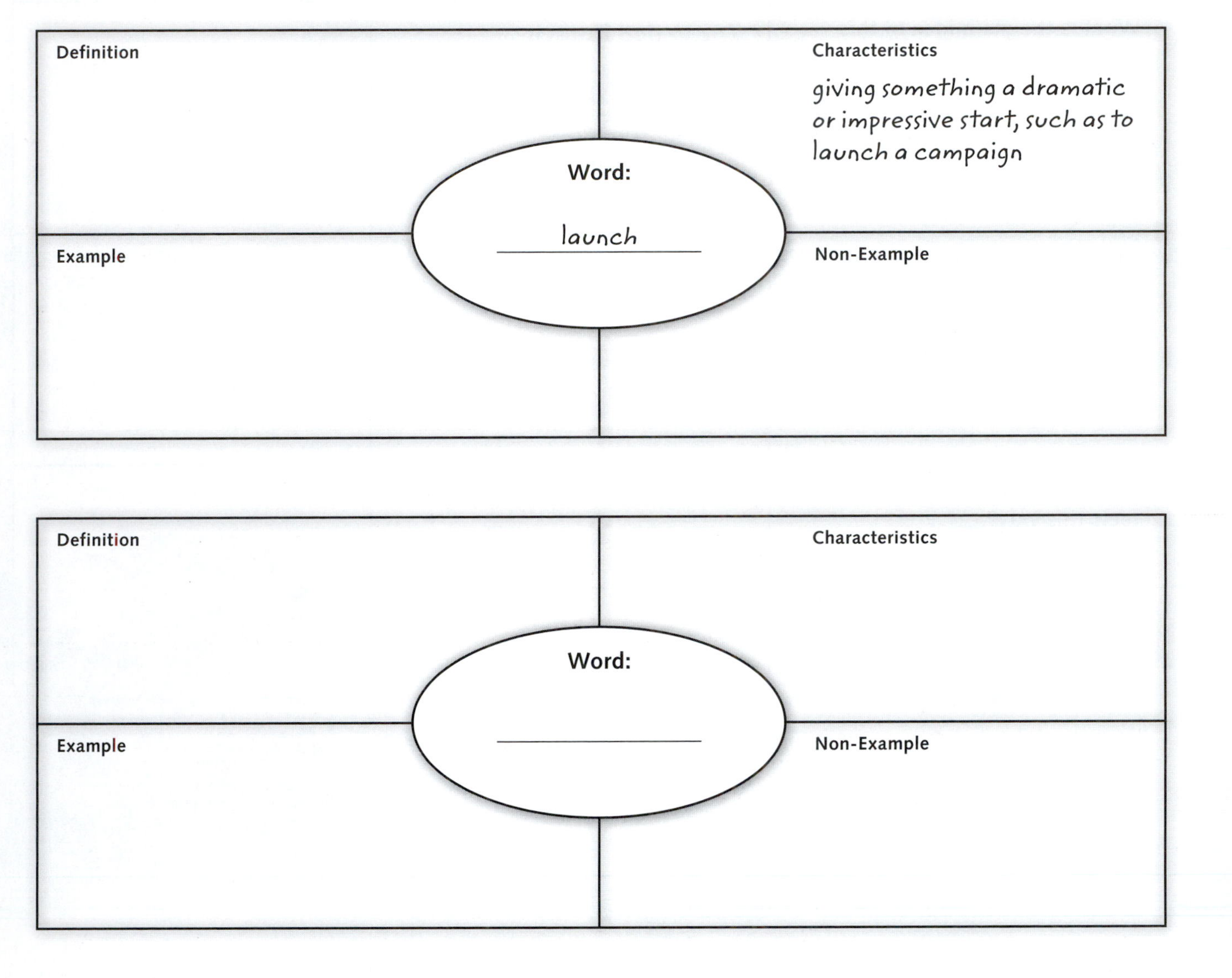

BLACK-AND-WHITE COPY READY

Name Class Date

SECTION 2 HISTORY
GeoActivity

Use with Southeast Asia Geography & History, Section 2.4, *in your textbook.*

Go to Interactive Whiteboard GeoActivities *at* **myNGconnect.com** *to complete this activity online.*

NATIONAL GEOGRAPHIC School Publishing

2.4 DOCUMENT-BASED QUESTION: THE VIETNAM WAR

Compare and Contrast Two Wars in Asia

The United States became involved in two wars in Asia in the decades following the end of World War II—the Korean War (1950–1953) between North Korea and South Korea and the Vietnam War (1954–1976) between North Vietnam and South Vietnam. The United States entered both conflicts to prevent the spread of communism in Asia.

Recall what you learned previously about the Korean War and the Korean Armistice Agreement. Then review the events listed below that led to the end of the war in Vietnam. Complete a Venn diagram to compare and contrast the way the two wars came to an end.

Events Leading to the End of the Vietnam War

- **1968:** As U.S. casualties increase, support for the war weakens and public opinion is divided. Antiwar protests spread.
- **1969:** President Nixon announces a "Vietnamization" plan to turn fighting over to the South Vietnamese. Withdrawal of U.S. troops begins.
- **1973:** Secret negotiations lead to the Paris Peace Accords between the United States and North Vietnam.
- **1975:** Saigon falls to the North Vietnamese army.
- **1976:** The country is united under North Vietnamese leader Ho Chi Minh.

1. **Create Graphic Organizers** Put events that led to the end of the Korean War in the left oval and events that led to the end of Vietnam War in the right oval. Write what was similar about the situations in the center area.

Korean War

International UN forces came to aid South Korea

Similarities

Vietnam War

2. **Compare and Contrast** What is the most important difference in the way these two wars ended? Think about the situation for each country after the war.

___

___

___

Name ____________________ Class ____________ Date ____________

SECTION 2 HISTORY

2.1–2.4 Review and Assessment

Use with Southeast Asia Geography & History, Sections 2.1–2.4, *in your textbook.*

Follow the instructions below to review what you have learned in this section.

Vocabulary Next to each vocabulary word, write the letter of the correct definition.

1.____ complex
2.____ bas-relief
3.____ monopoly
4.____ colonialism
5.____ fossil
6.____ commerce
7.____ launch
8.____ resistance

A. the business of trading goods and services
B. a set of interconnected buildings
C. the preserved remains or impression of a life form from an earlier geologic age
D. the complete control of a market
E. opposition, or a group that rebels or opposes
F. a sculpture that slightly projects from a flat background
G. to start something or set it in motion
H. one country ruling and developing trade in another country for its own benefit

Main Ideas Use what you've learned about Southeast Asia's history to answer these questions.

9. **Describe Geographic Information** Why did the waterways through Southeast Asia become important to trade?

____

10. **Compare and Contrast** How did the social standing of women in ancient Vietnam compare with that of women in China?

____

11. **Sequence Events** Which empire in modern-day Indonesia formed first?

____

12. **Explain** What do the different levels of the Borobudur temple complex represent?

____

13. **Identify** What organization from the Netherlands dominated trade in Southeast Asia for many years?

____

14. **Compare and Contrast** How did Thailand and parts of the Philippines differ from the rest of the region in the 19th century?

____

15. **Explain** Why do experts believe Indonesia may have been a home of early humans?

____

16. **Describe** How did Dutch control over Indonesia change during the 19th century?

____

17. **Analyze Cause and Effect** How did Spain's loss of economic power in the 1800s affect life for Filipinos?

____

18. **Explain** Why did the United States support South Vietnam during the Vietnam War?

____

Focus Skill: Make Inferences

Answer the questions below to make inferences about Southeast Asia and the ways that the region's geography has influenced its history.

19. Why do you think China sought to conquer Vietnam?

____

____

20. Did India or China have a stronger influence on the culture of the Khmer Empire? Why?

____

____

21. How do you think the Srivijaya Empire's control of the Strait of Malacca affected the kingdom's wealth?

____

____

22. Were spices such as cinnamon, nutmeg, and black pepper plentiful in Europe in the 1500s? Explain.

____

____

23. Why do you think the Indonesian people were able to gain their independence in the late 1940s?

____

____

24. In what way did colonialism eventually strengthen Indonesian national unity?

____

____

25. Why did Japan attack the Philippines in 1941?

____

____

26. South Vietnam surrendered two years after U.S. forces withdrew. What inferences can be drawn about the prospects for success at the time the United States left?

____

____

Synthesize: Answer the Essential Question

How have physical barriers in Southeast Asia influenced its history? Recall what you have read about the importance of the region's waterways to trade, as well as the flourishing of different kingdoms on the region's many islands. Be sure to discuss ways that the region's physical geography affected its history during centuries of colonialism, as well as the region's military importance during the 20th century.

____

____

____

____

____

____

____

SECTION 2 HISTORY

2.1–2.4 Standardized Test Practice

Use with Southeast Asia Geography & History, Sections 2.1–2.4, *in your textbook.*

Follow the instructions below to practice test-taking on what you've learned from this section.

Multiple Choice Circle the best answer for each question from the options available.

1. Chinese culture arrived in the region when the Chinese invaded which modern-day country?

A Cambodia
B Vietnam
C Indonesia
D Thailand

2. Angkor Wat was dedicated to what deity?

A Vishnu
B Buddha
C Allah
D Krishna

3. What was the effect on Angkor Wat after forces from modern-day Thailand conquered the city?

A The complex was converted into a Buddhist temple.
B The complex was restored to its original condition.
C The complex was destroyed to demonstrate the army's power.
D The complex fell into ruin.

4. How did the economies of Southeast Asian kingdoms change under colonialism?

A imports increased
B imports decreased
C exports increased
D exports decreased

5. In what century did colonialism in Southeast Asia end?

A the 17th century
B the 18th century
C the 19th century
D the 20th century

6. How did control over the Philippines change in 1898?

A Control transferred from Spain to the United States.
B Great Britain defeated France and took control.
C Control transferred from Spain to the indigenous Filipino population.
D The Netherlands and Portugal agreed to divide the Philippines between them.

7. In colonial times, what did Europeans call parts of Indonesia?

A the Gold Islands
B the Spice Islands
C the Silk Islands
D the Ivory Islands

8. Who was Emilio Aguinaldo?

A an explorer who claimed the Philippines for Spain
B the founder of Manila
C a 19th-century Spanish governor in the Philippines
D a leader in the Filipino independence movement

9. What action started the Vietnam War in 1959?

A North Vietnam attempted to overthrow South Vietnam's government.
B South Vietnam claimed North Vietnamese territory.
C The United States offered military support for North Vietnam.
D North Vietnam invaded Cambodia.

10. After South Vietnam surrendered to North Vietnam in 1975, what was the reunited country called?

A the Democratic People's Republic of Vietnam
B the Socialist Republic of Vietnam
C the Communist Federation of Vietnam
D the Kingdom of Vietnam

Document-Based Question

The following article, **"Close Encounters at a Khmer Temple"** by Douglas Preston, appeared in *National Geographic* in August 2000. Read the passage and answer the questions below.

> *We parked on the outskirts in the shade of a gum tree and stretched our aching bodies. The ruin of Banteay Chhmar lay a quarter mile distant, concealed in the forest. We hiked across fields and forced our way into the undergrowth. Soon a massive wall of greenish sandstone, covered with spectacular bas-reliefs, loomed up ahead. The enormous temple—it covers more than 500,000 square feet [46,500 square meters]—was submerged under giant silk-cotton trees, thick bushes, banyans dropping curtains of vines, and rank, steaming vegetation.*

Constructed Response

Read each question carefully and write your answer in the space provided.

11. Preston writes that the temple was "concealed in the forest." Based on the context, what do you think *concealed* means?

____

12. Do you think the temple was meant to be concealed when it was built? Why or why not?

____

Extended Response

Read each question carefully and write your answer in the space provided.

13. The passage does not identify the temple's location. What country do you think Banteay Chhmar is in? What religion do you think was meant to be honored with the temple's construction? Explain your answer.

____

____

Angkor Wat

14. How would you describe the architectural style of Angkor Wat?

____

____

15. In addition to the stone structures shown in the photo, many other parts of Angkor Wat were originally built out of wood. Why do you think these wooden structures have not survived?

____

____

Southeast Asia

TODAY • RESOURCE BANK

Name Class Date

SECTION 1 CULTURE

1.1 Religious Traditions

Use with Southeast Asia Today, Section 1.1, *in your textbook.*

Reading and Note-Taking Categorize Religions

As you read Section 1.1, use the Classification Chart below to help you keep track of different religions in Southeast Asia. Take notes about a separate religion in each box.

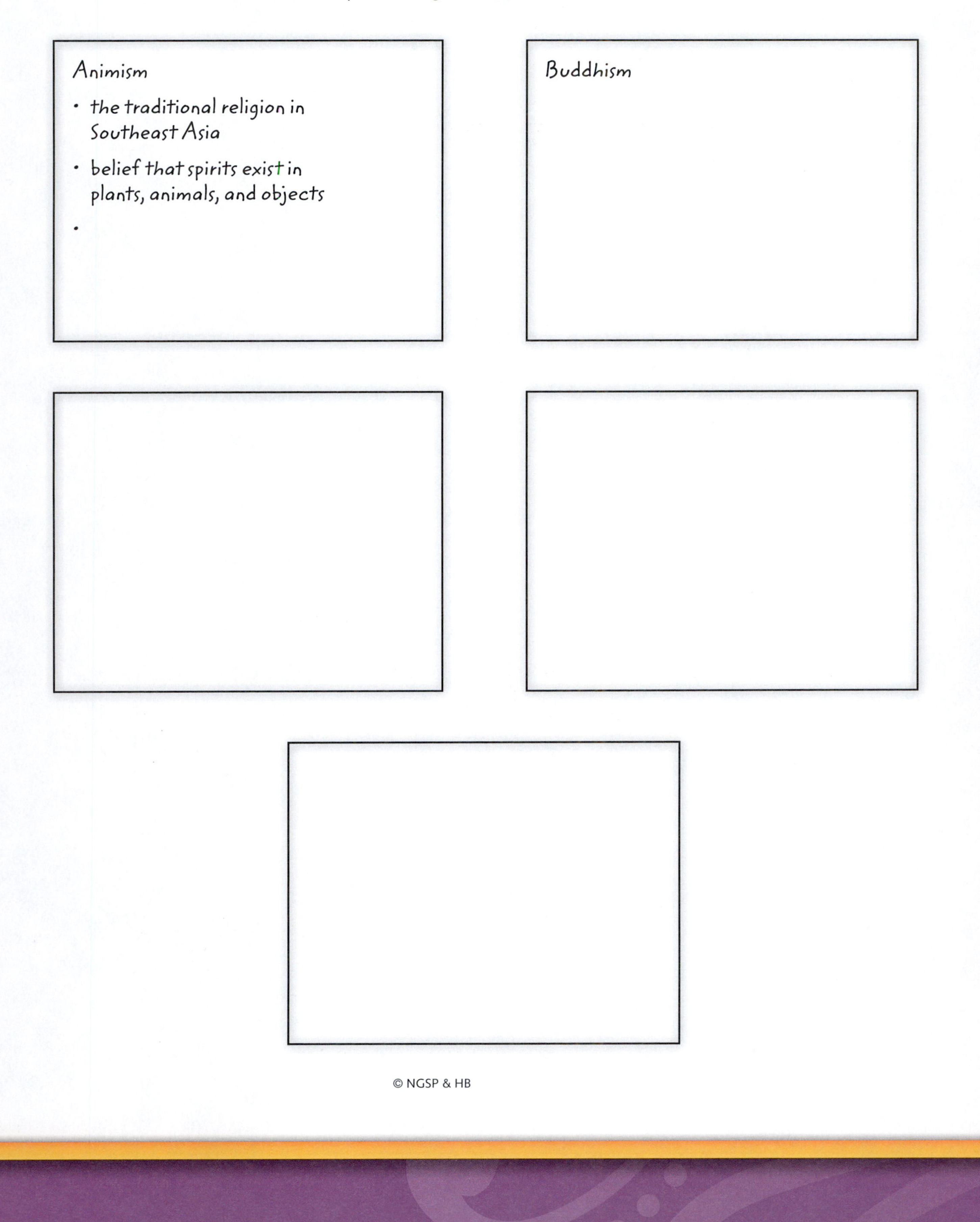

BLACK-AND-WHITE COPY READY

BLACK-AND-WHITE COPY READY

Name Class Date

SECTION 1 CULTURE

1.1 Religious Traditions

Use with Southeast Asia Today, Section 1.1, *in your textbook.*

Vocabulary Practice

KEY VOCABULARY

- **prehistoric** (pree-hihs-TAWR-ihk) adj., existing or occurring before written history
- **ritual** (RIH-chuh-uhl) n., a formal and regularly repeated action, such as a religious ceremony

ACADEMIC VOCABULARY

- **predominant** (prih-DAH-muh-nuhnt) adj., more common or more important than others

Write a sentence about culture in your region using the Academic Vocabulary word *predominant*.

____

____

Word Map Complete a Word Map for the Key Vocabulary words *prehistoric* and *ritual*. Write what the word means in your own words. To describe what it is like, provide one or two related words or synonyms.

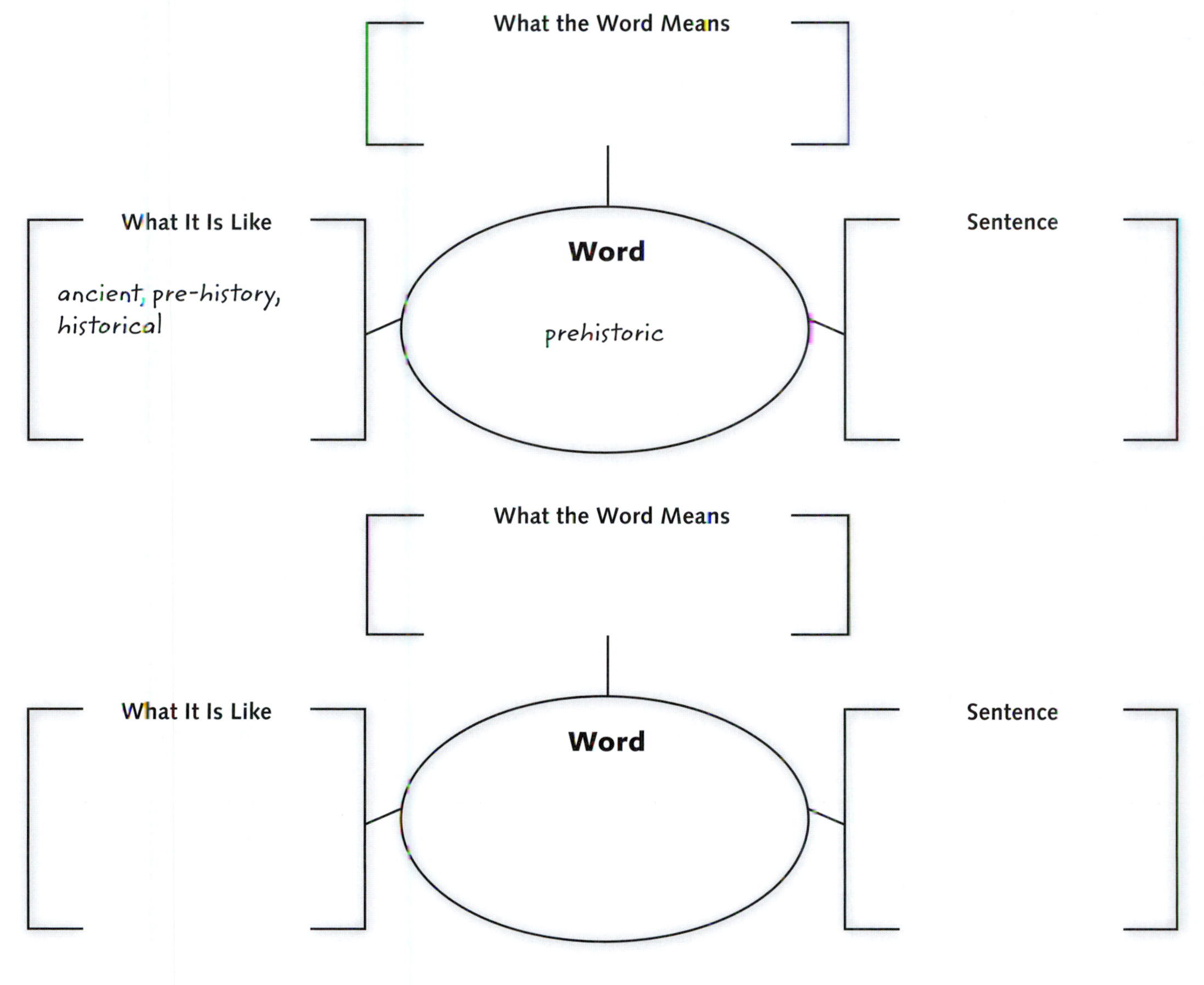

Name ______ Class ______ Date ______

SECTION 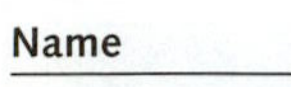 CULTURE

GeoActivity

Use with Southeast Asia Today, Section 1.1, *in your textbook.*

Go to Interactive Whiteboard GeoActivities *at* **myNGconnect.com** *to complete this activity online.*

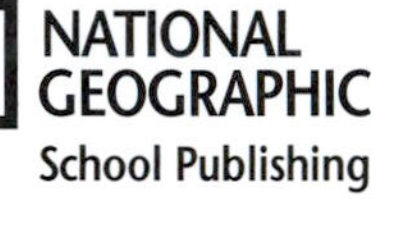

1.1 RELIGIOUS TRADITIONS

Map Religion in Southeast Asia

As you have learned, a variety of religions are practiced today in Southeast Asia. The chart below lists the distribution of religions in each country in the region. Study the chart and then answer the questions that follow.

Major Religious Populations in Southeast Asia
(Percentage of Total Population)

COUNTRY	BUDDHISM	CHRISTIANITY	ISLAM	OTHER/NONE
Brunei	13%	10%	67%	10%
Cambodia	96%	*	2%	2%
Indonesia	*	9%	86%	5%
Laos	67%	2%	*	31%
Malaysia	19%	9%	60%	11%
Myanmar	89%	4%	4%	3%
Philippines	*	93%	5%	3%
Singapore	43%	15%	15%	28%
Thailand	95%	1%	5%	0%
Timor-Leste	0%	99%	1%	0%
Vietnam	9%	7%	*	83%

Source: CIA World Factbook, 2011

*Included in "Other/None" for this country

1. **Create Maps** Choose a different color for each religion in the chart and add it to the legend on the map at right. Shade each country on the map according to its dominant, or most popular, religion.

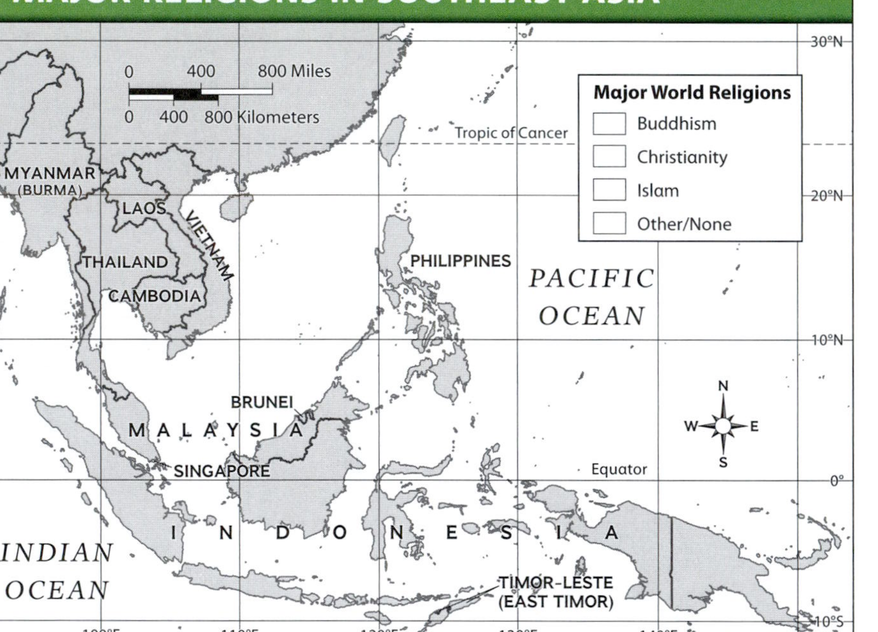

2. **Analyze Data** Which countries have a strong religious majority (80 percent or more) and what are the religions?

______

______

3. **Synthesize** The majority of people in Vietnam claim to follow no religion. Based on what you know about the country's history and government, what could be a possible explanation for this trend?

______

BLACK-AND-WHITE COPY READY

Name Class Date

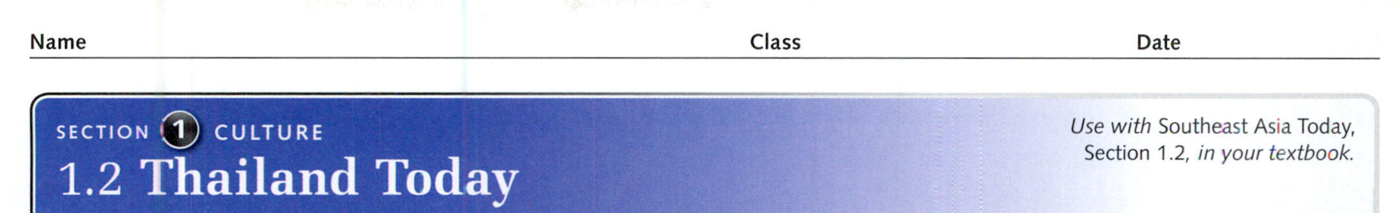

SECTION 1 CULTURE

1.2 Thailand Today

Use with Southeast Asia Today, Section 1.2, *in your textbook.*

Reading and Note-Taking Summarize Details

Use the Detail Web below to record information about Thai culture as you read Section 1.2. Then summarize the details in your web to answer the question.

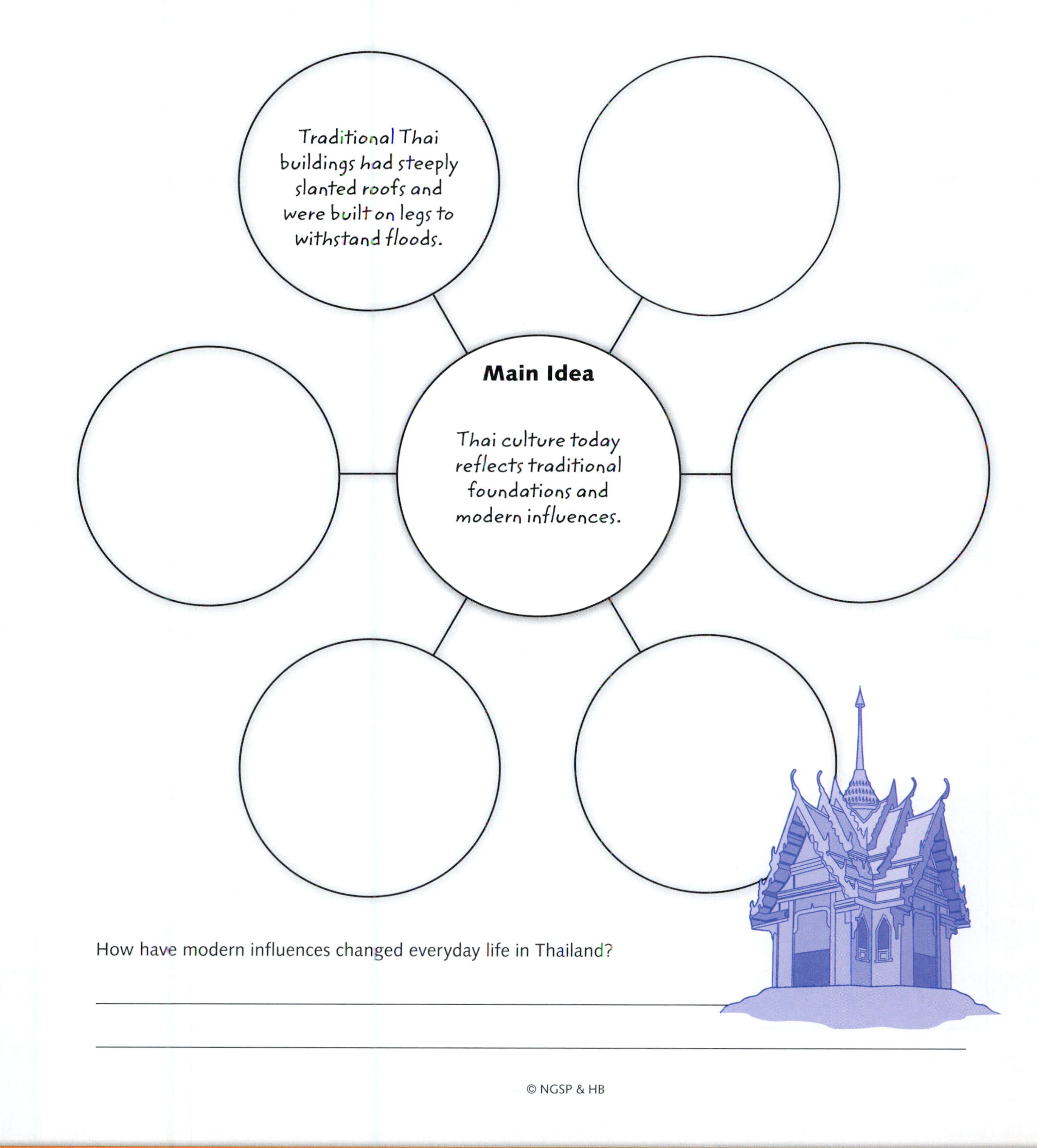

How have modern influences changed everyday life in Thailand?

Name Class Date

SECTION 1 CULTURE
1.2 Thailand Today

Use with Southeast Asia Today, Section 1.2, *in your textbook.*

Vocabulary Practice

KEY VOCABULARY

- **attribute** (A-truh-byoot) n., a specific quality or feature
- **metropolitan area** (meh-truh-PAH-luh-tuhn EHR-ee-uh) n., the populated location that includes a city and the surrounding territory
- **monk** (MUHNGK) n., a man who devotes himself to religious work
- **wat** (WAHT) n., Buddhist temple

Definition Chart Complete a Definition Chart for the Key Vocabulary words *attribute* and *metropolitan area.* Then create Definition Charts on your own paper for *monk* and *wat.*

Word	attribute
Definition	a specific quality or feature
In Your Own Words	
Symbol or Diagram	

Word	metropolitan area
Definition	
In Your Own Words	
Symbol or Diagram	

BLACK-AND-WHITE COPY READY

Name Class Date

SECTION 1 CULTURE

GeoActivity

Use with Southeast Asia Today, Section 1.2, *in your textbook.*

Go to Interactive Whiteboard GeoActivities *at* **myNGconnect.com** *to complete this activity online.*

NATIONAL GEOGRAPHIC School Publishing

1.2 THAILAND TODAY

Graph Thailand's Population Trends

Thailand is the largest rice exporter in the world, yet agriculture makes up only 12 percent of the country's GDP. During the last half of the 20th century, Thailand's economy shifted from a reliance on agriculture to a reliance on manufacturing. What happened to Thailand's population during this time? The following chart shows Thailand's rural, urban, and total populations over the last 50 years. Study the chart and then answer the questions.

Thailand's Population, 1969-2009

	RURAL POPULATION	URBAN POPULATION	TOTAL POPULATION
1969	28,664,594	7,509,805	36,174,399
1979	34,116,350	12,111,767	46,228,117
1989	39,616,617	16,291,677	55,908,294
1999	42,695,651	19,128,344	61,823,995
2009	44,954,659	22,809,374	67,764,033

1. **Analyze Data** What trend do you notice about all three populations (rural, urban, and total) between 1969 and 2009?

2. **Calculate** Calculate the percentage of the total population that was rural and urban for each year. To figure out the percentage, divide the rural or urban population by the total population and multiply by 100. Write the percentage next to each amount in the chart above.

3. **Create Graphs** Create a bar graph for the percentage of Thailand's rural and urban populations between 1969 and 2009 in the space below. Use a different color for each population and indicate them in the legend.

Rural and Urban Populations, 1969-2009

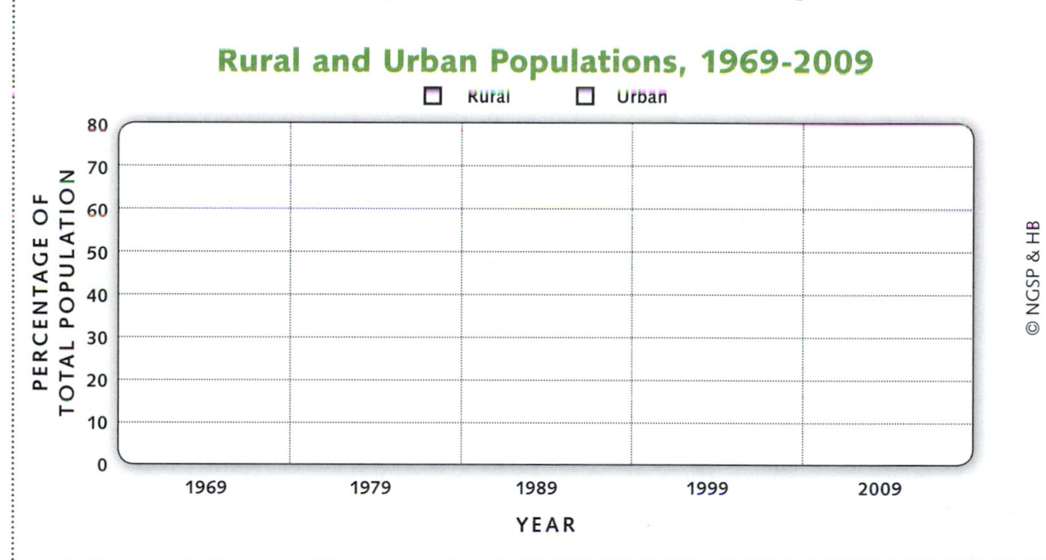

4. **Interpret Graphs** What trend do you notice about the percentage of rural and urban populations between 1969 and 2009?

5. **Make Inferences** How might Thailand's changing economy have affected both the rural and urban populations?

SECTION 1 CULTURE

1.3 Regional Languages

Use with Southeast Asia Today, Section 1.3, *in your textbook.*

Reading and Note-Taking Create a Chart

Organize information in a Three-Column Chart as you read about the different languages spoken across Southeast Asia in Section 1.3. Give examples from the text and the Spoken Languages infographic.

Official Languages	Dialects	Non-native Languages
• An official language is the dominant native language. • It's used in government, business, education, and the media.		

BLACK-AND-WHITE COPY READY

BLACK-AND-WHITE COPY READY

Name Class Date

SECTION 1 CULTURE

1.3 Regional Languages

Use with Southeast Asia Today, Section 1.3, *in your textbook.*

Vocabulary Practice

KEY VOCABULARY

- **adapt** (uh-DAPT) v., to adjust to new conditions
- **dialect** (DY-uh-lehkt) n., a regional variation of a main language
- **language diffusion** (LANG-gwihj duh-FYOO-shuhn) n., the spread of a language from its original home

Word Squares Complete the Word Squares below for the Key Vocabulary words *adapt* and *dialect*. On separate paper, create a Word Square for *language diffusion*.

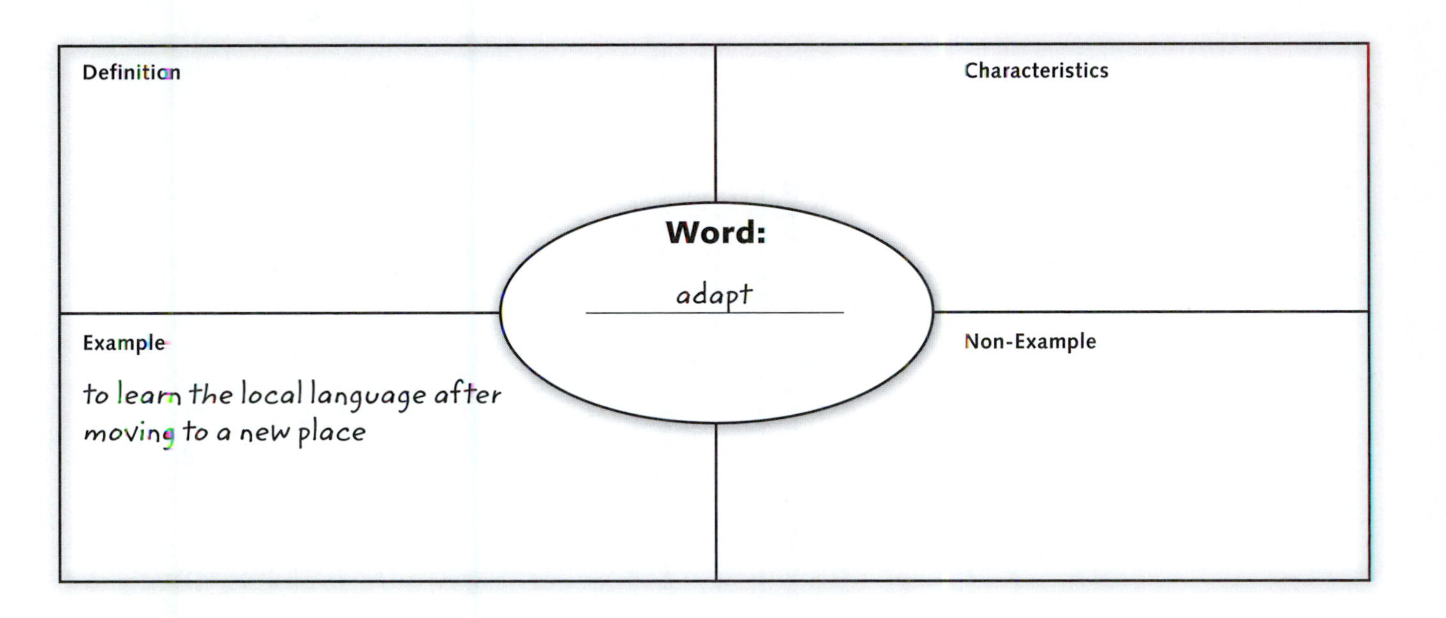

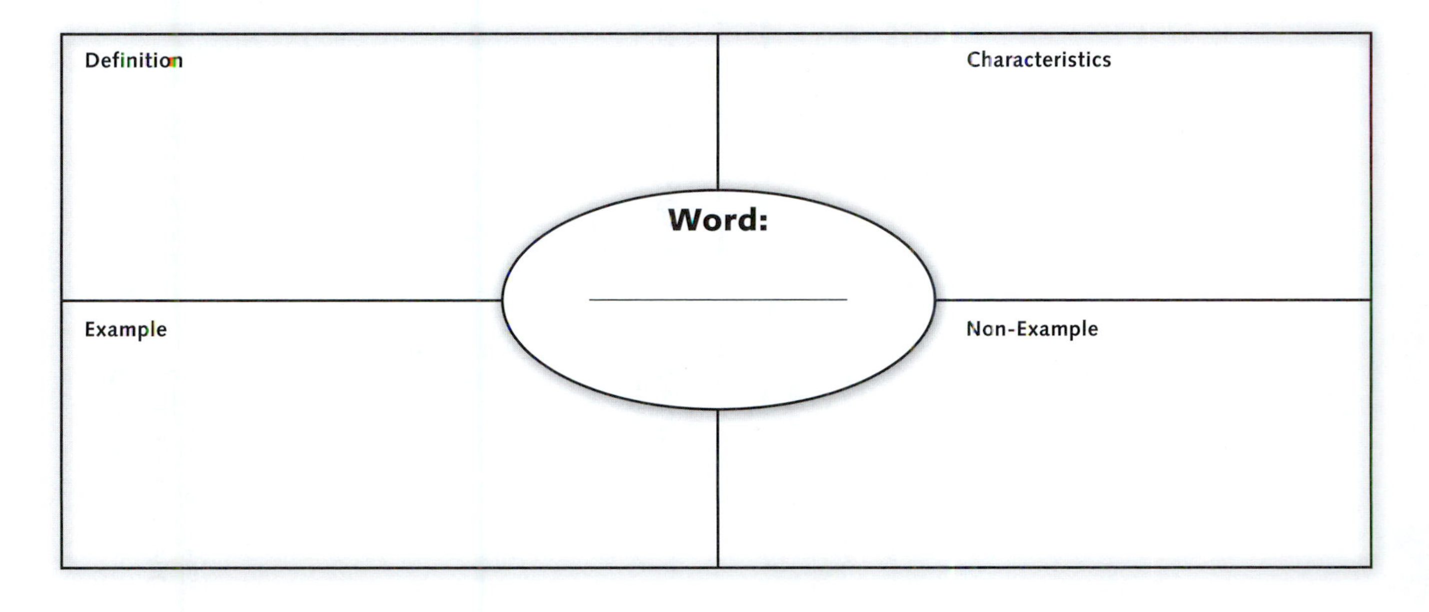

Name ______ Class ______ Date ______

SECTION 1 CULTURE
GeoActivity

Use with Southeast Asia Today, Section 1.3, *in your textbook.*
Go to Interactive Whiteboard GeoActivities *at* **myNGconnect.com** *to complete this activity online.*

NATIONAL GEOGRAPHIC School Publishing

1.3 REGIONAL LANGUAGES

Analyze Language Relationships

Linguistics is the study of language. Linguists are people who study how languages developed over time. In order to understand the relationships among languages, linguists categorize them in language families. A language family is a group of languages that developed from a common ancestral, or early, language. A language family is often displayed in the form of a language tree. The language trees at right show the roots of several common languages spoken in Southeast Asia. Study the trees and then answer the questions.

1. **Interpret Graphics** Which languages are more closely related—Bahasa Malaysia and Bahasa Indonesian or Khmer and Vietnamese? How can you tell?

______

______

______

2. **Draw Conclusions** How might these different languages have developed from a common ancestral language? Think about the geography of the region as you formulate your response.

______

______

______

Selected Southeast Asian Languages

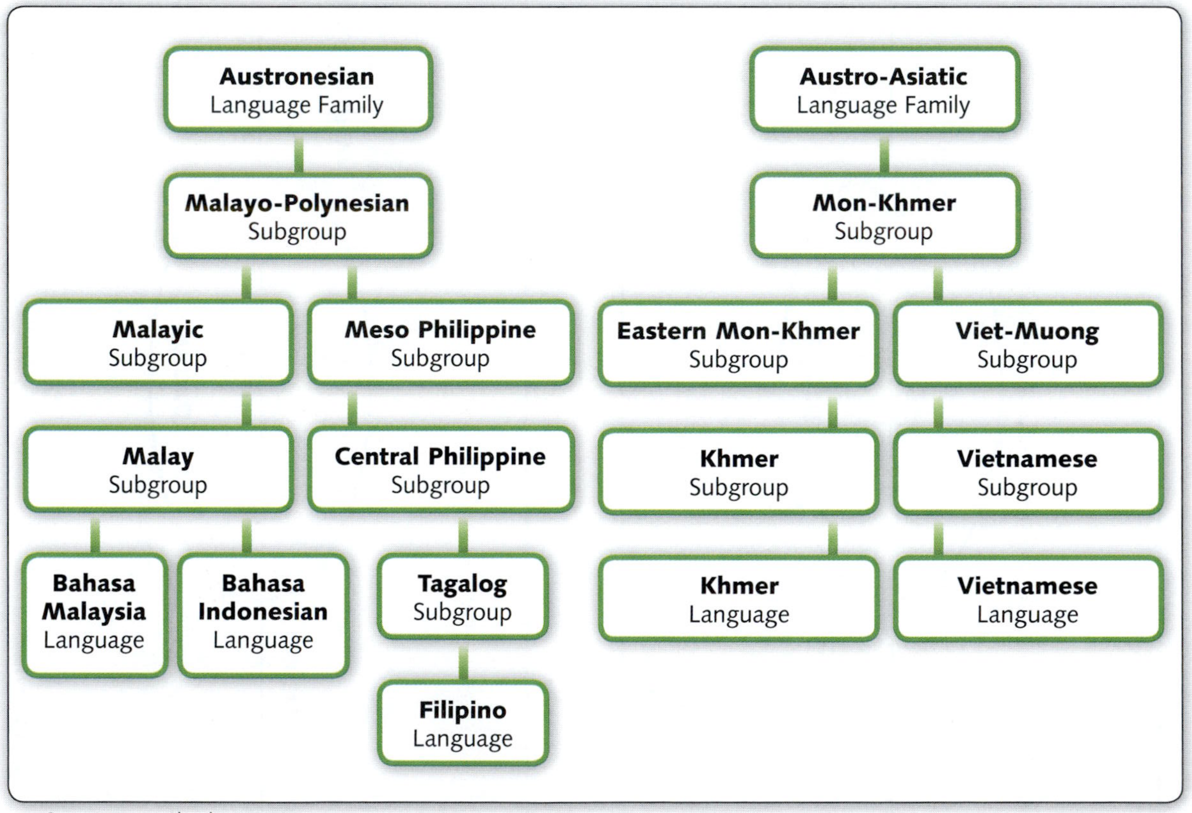

- **Austronesian** Language Family
  - **Malayo-Polynesian** Subgroup
    - **Malayic** Subgroup
      - **Malay** Subgroup
        - **Bahasa Malaysia** Language
        - **Bahasa Indonesian** Language
    - **Meso Philippine** Subgroup
      - **Central Philippine** Subgroup
        - **Tagalog** Subgroup
          - **Filipino** Language
- **Austro-Asiatic** Language Family
  - **Mon-Khmer** Subgroup
    - **Eastern Mon-Khmer** Subgroup
      - **Khmer** Subgroup
        - **Khmer** Language
    - **Viet-Muong** Subgroup
      - **Vietnamese** Subgroup
        - **Vietnamese** Language

Source: www.ethnologue.com

Name Class Date

SECTION 1 CULTURE

1.4 Saving the Elephant

Use with Southeast Asia Today, Section 1.4, *in your textbook.*

Reading and Note-Taking Analyze Cause and Effect

As you read Section 1.4, use the Cause-and-Effect Chart below to help you understand the threats to Asian elephants.

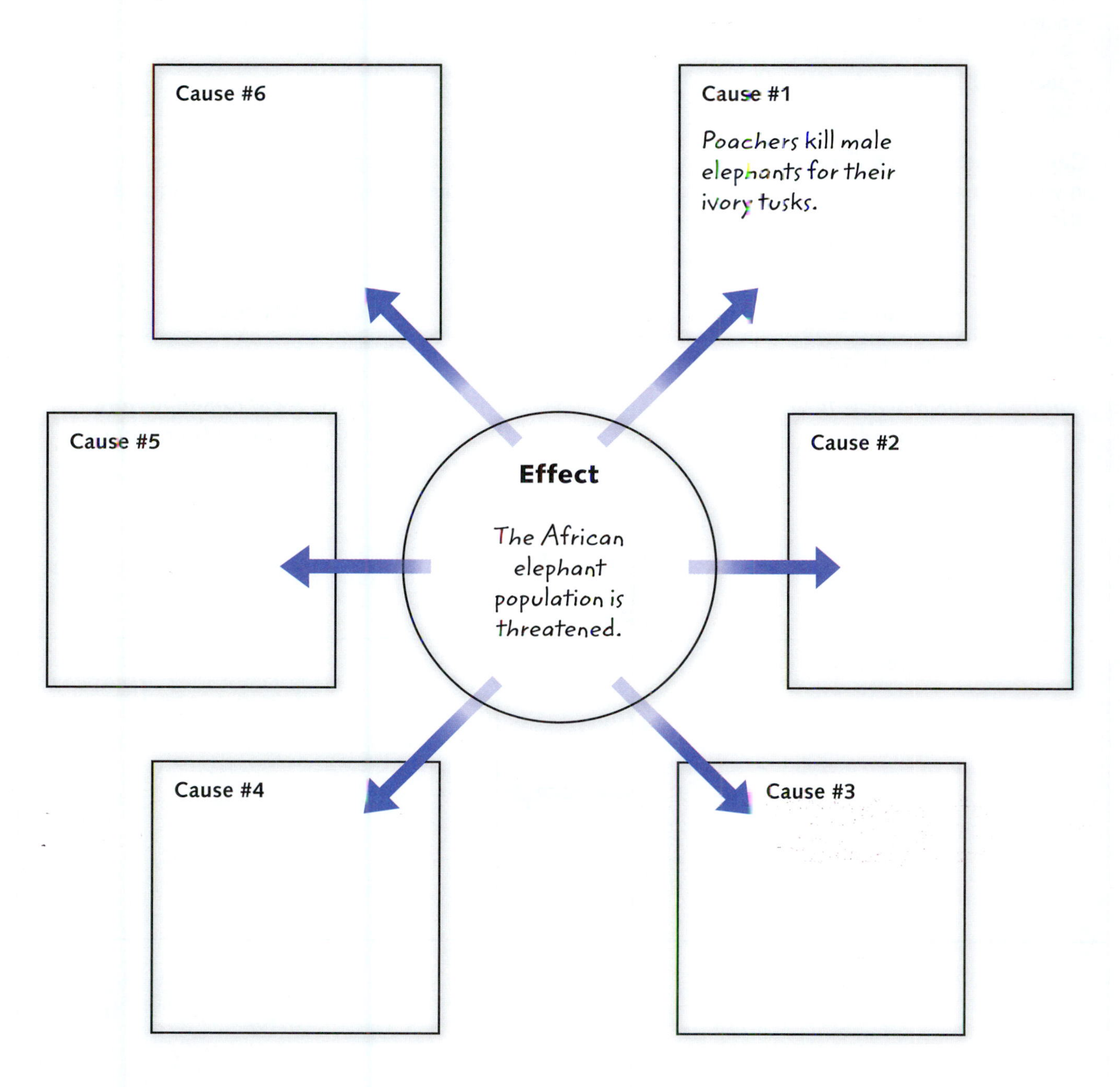

Identify Problems and Solutions Why is it important to minimize interactions between humans and elephants in the wild?

Name Class Date

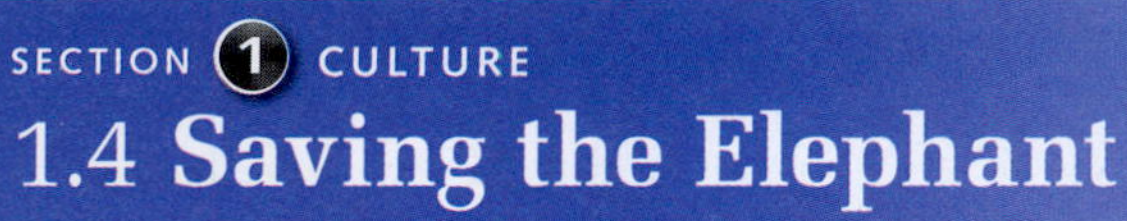

SECTION 1 CULTURE

1.4 Saving the Elephant

Use with Southeast Asia Today, Section 1.4, *in your textbook.*

BLACK-AND-WHITE COPY READY

Vocabulary Practice

KEY VOCABULARY

- **domesticate** (duh-MEHS-tih-kayt) v., to tame an animal or train it to work with humans
- **poach** (POHCH) v., to hunt a wild animal illegally
- **restore** (rih-STOHR) v., to return something to its earlier, better condition

Cause and Effect Chart For each version of the Key Vocabulary word below, provide a definition in your own words. In the Effect column for each word, write a sentence explaining how the human action affects Asian elephants.

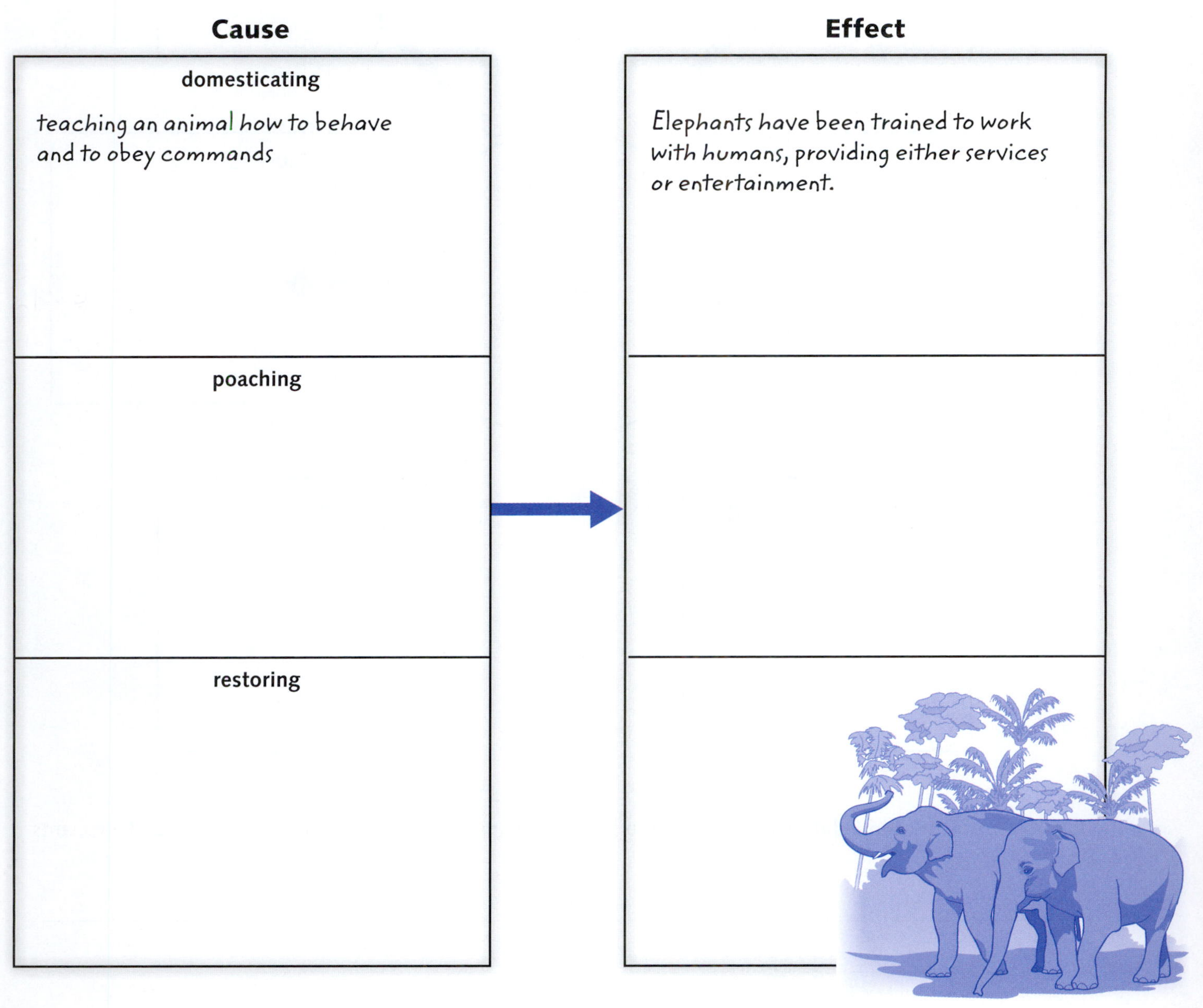

Cause	Effect
domesticating *teaching an animal how to behave and to obey commands*	*Elephants have been trained to work with humans, providing either services or entertainment.*
poaching	
restoring	

Name ______________________ Class ______________ Date ______________

SECTION 1 CULTURE
GeoActivity

Use with Southeast Asia Today, Section 1.4, *in your textbook.*
Go to Interactive Whiteboard GeoActivities *at* **myNGconnect.com** *to complete this activity online.*

NATIONAL GEOGRAPHIC School Publishing

1.4 SAVING THE ELEPHANT

Investigate Endangered Species

You have read about some of the ways Southeast Asia's conservationists work to protect the region's elephants. How do these methods compare with efforts to protect other endangered species? Use the research links at **Connect to NG** to research threats to some of the region's rare species and how people are trying to protect them. Use your research information to complete the graphic organizer at right. Compare and contrast one animal from your research with the Asian elephant. Then answer the question.

Draw Conclusions Why is it important for governments to work with each other in order to protect endangered species?

____

____

____

____

____

____

____

____

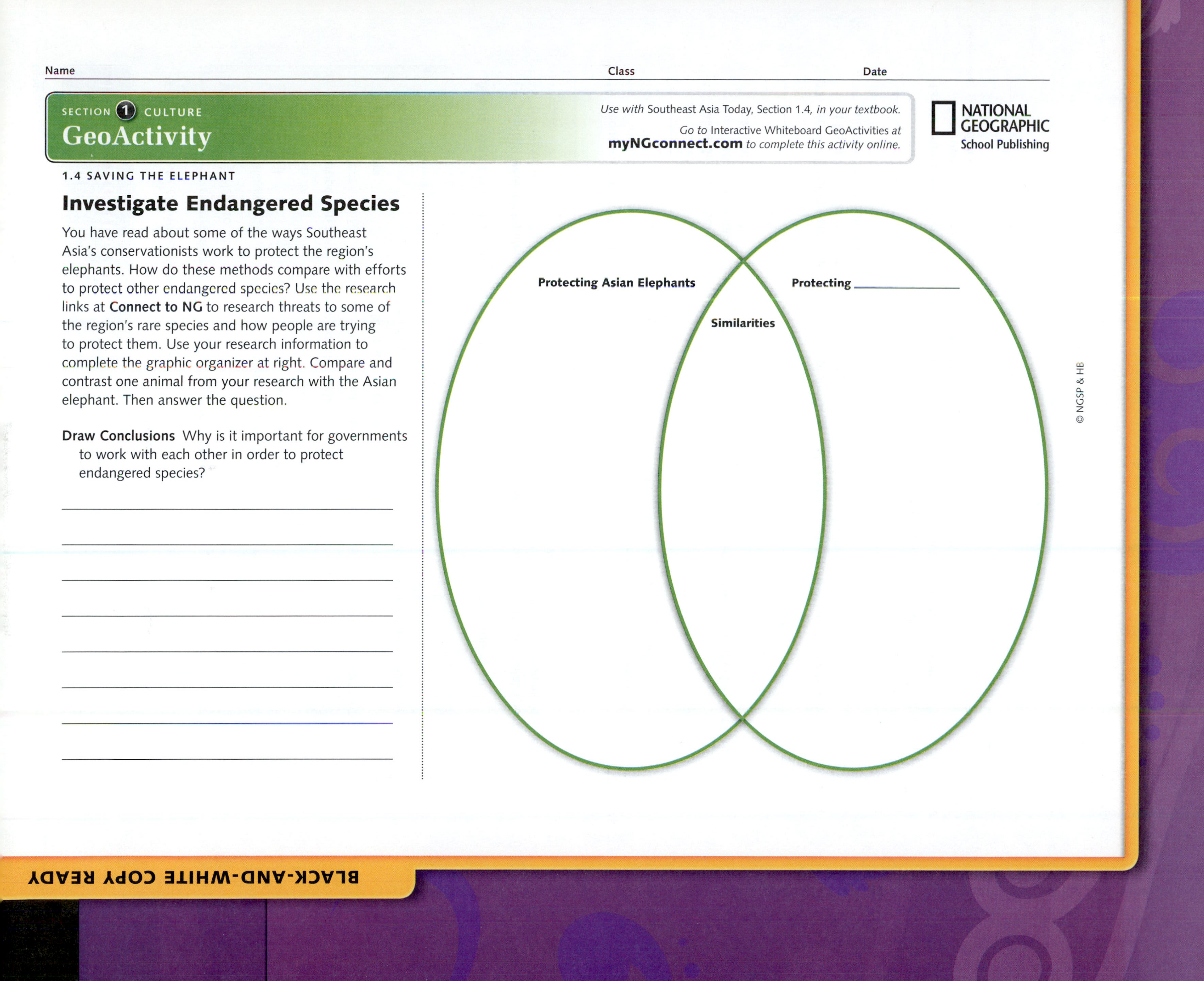

BLACK-AND-WHITE COPY READY

Name Class Date

SECTION 1 CULTURE

1.1–1.4 Review and Assessment

Use with Southeast Asia Today, Sections 1.1–1.4, *in your textbook.*

Follow the instructions below to review what you have learned in this section.

Vocabulary Next to each vocabulary word, write the letter of the correct definition.

1.____ prehistoric
2.____ ritual
3.____ attribute
4.____ metropolitan area
5.____ dialect
6.____ adapt
7.____ restore
8.____ domesticate

A. to return something to its earlier, better condition
B. to adjust to new conditions
C. existing or occurring before written history
D. a regional variation of a main language
E. a formal and regularly repeated action, such as a religious ceremony
F. to tame an animal or train it to work with humans
G. the populated location that includes a city and the surrounding territory
H. a specific quality or feature

Main Ideas Use what you've learned about Southeast Asia's culture to answer these questions.

9. **Explain** How did Buddhism first come to the region?

____

10. **Sequence Events** Which religion arrived in Southeast Asia earlier—Islam or Christianity?

____

11. **Make Generalizations** What religion is the most prominent today in Southeast Asia's mainland countries?

____

12. **Identify Problems and Solutions** What feature of traditional Thai architecture helps buildings withstand monsoon floods?

____

13. **Describe Geographic Information** Four out of five young people now work in what kind of location in Thailand?

____

14. **Make Inferences** Why do you think speakers of different dialects often live in small, isolated communities?

____

15. **Explain** Why was the Malay language important to regional trade for centuries?

____

16. **Summarize** How does trade affect a country's languages?

____

17. **Explain** Why have Asian elephants lost habitat in recent years?

____

18. **Identify Problems and Solutions** What is one way modern technology is used to protect Asian elephants?

____

BLACK-AND-WHITE COPY READY

BLACK-AND-WHITE COPY READY

Focus Skill: Analyze Cause and Effect

Answer the questions below to analyze the ways local traditions and outside influences have shaped cultures in Southeast Asia.

19. What effect do animists believe their rituals will have?

____

____

20. Over the centuries, what was the main reason a country's predominant religion changed?

____

____

21. How did the Indian, Khmer, and Chinese cultures influence Thai architecture?

____

____

22. What is one way urban life has affected Thai culture?

____

____

23. How has Myanmar's diversity of languages affected national unity?

____

____

24. Why is English among the languages spoken in many Southeast Asian countries?

____

____

25. How does human population growth affect Asian elephants?

____

____

26. How does hanging hammocks near crops affect elephants' behavior? Why?

____

____

Synthesize: Answer the Essential Question

How have local traditions and outside influences shaped cultures in Southeast Asia? Consider the different factors that have produced the region's religious diversity and its variety of languages. Recall as well what you have read about the balance between tradition and modernity in Thailand.

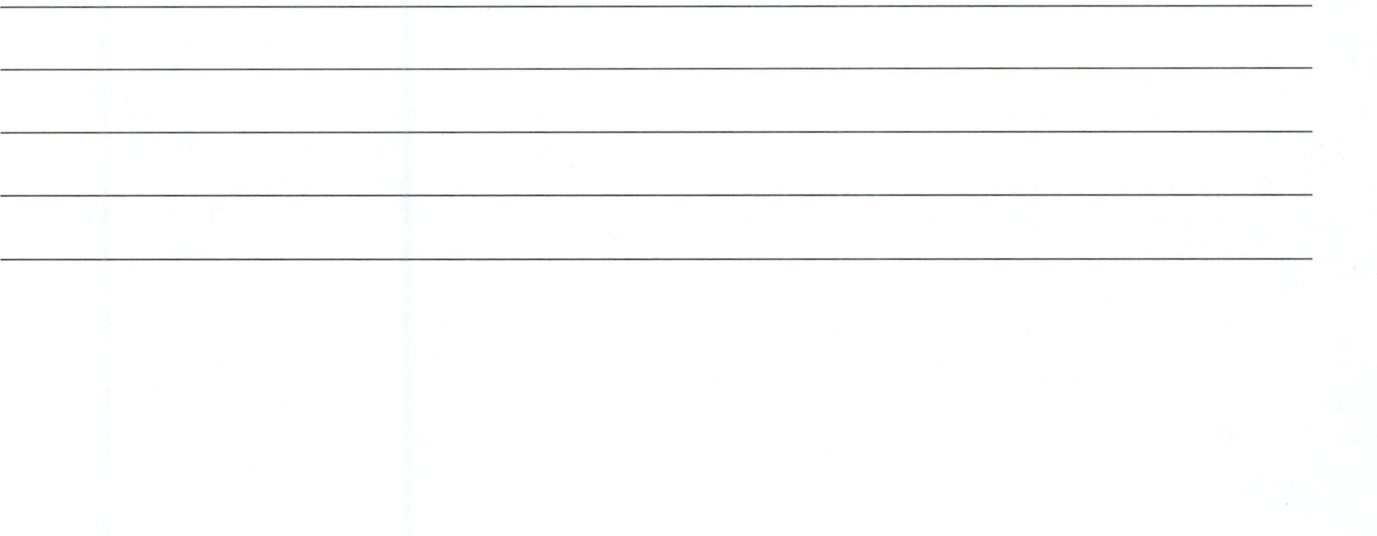

SECTION 1 CULTURE

1.1–1.4 Standardized Test Practice

Use with Southeast Asia Today, Sections 1.1–1.4, *in your textbook.*

Follow the instructions below to practice test-taking on what you've learned from this section.

Multiple Choice Circle the best answer for each question from the options available.

1. Traders and pilgrims from what country brought Buddhism and Hinduism to the region?

A China
B India
C Japan
D Taiwan

2. The major religion in the Philippines and East Timor shows the influence of what two European countries?

A England and France
B Spain and Portugal
C Germany and Switzerland
D Belgium and the Netherlands

3. Why do traditional Thai buildings have steeply slanted roofs?

A The roofs were designed not to collapse under heavy snow.
B The roofs help circulate air through homes during hot summers.
C The roofs were designed to shed heavy monsoon rains.
D The roofs were built to withstand heavy winds during typhoons.

4. Young men in Thailand traditionally commit to becoming monks for how long?

A at least three months
B between one and three years
C five years
D more than ten years

5. How has the movement of young people away from rural communities affected religion in Thailand?

A Monasteries in Thailand's cities have grown.
B Fewer young men are becoming monks.
C Young people of both genders are becoming more religious.
D Monks are devoting themselves to longer religious commitments.

6. What effect did having a common language have on Malaysia and Indonesia?

A It created conflict about which language would become the official language.
B It helped local populations rebel against colonial powers.
C It helped unify areas that are geographically separate.
D It made trade with people from other language regions more difficult.

7. In addition to the local official languages, French is commonly spoken in which two countries?

A Thailand and Myanmar
B Cambodia and Vietnam
C Indonesia and Malaysia
D Singapore and the Philippines

8. As Southeast Asian countries gained independence from colonial powers, what was the effect on Western languages in the region?

A Western languages remained the official languages in most countries.
B Western languages were banned.
C French and Dutch remained widely spoken, but all others were banned.
D Western languages became second languages.

9. Which of the following is true?

A From 2005 to 2010, no elephants were killed in Cambodia.
B Elephants are not native to Cambodia.
C Increased elephant populations in Cambodia threaten human habitat.
D From 2001 to 2008, poaching decreased Cambodia's elephant population.

10. Which statement best reflects the outcome of the agreement in 1975 that banned trade in ivory?

A Poaching of male Asian elephants for their ivory tusks has ceased.
B The Asian elephant population has doubled.
C Ivory is no longer valued.
D Illegal trade continues in spite of the ban.

Document-Based Question

The following article, **"Facing Down the Fanatics"** by Michael Finkel, appeared in *National Geographic* in October 2009. Read the passage and answer questions below.

> *Indonesia is tucked away in a far corner of the world map, a rain [sic] of islands just north of Australia, yet violence here can have global repercussions. It is the most populous Muslim country in the world, home to 207 million Muslims—36 million more than the next largest Muslim nation, Pakistan, and two-thirds as many as all the countries of the Middle East combined. It is extremely devout; a recent Pew Global Attitudes survey found that Indonesia was one of the world's most religious nations. It's also a thriving democracy, the third largest in the world, after India and the United States.*

Constructed Response Read each question carefully and write your answer in the space provided.

11. Why would events in Indonesia have a strong effect on the Muslim world in general?

12. What geographic factor helps explain why Finkel needs to describe basic facts about Indonesia's population?

Extended Response Read each question carefully and write your answer in the space provided.

13. Recall what you have read about religions in Southeast Asia. Why might religious-based conflict be a concern in Indonesia and other countries in the region?

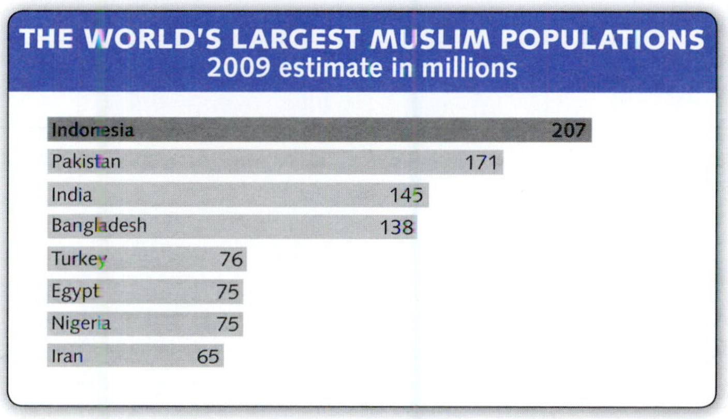

THE WORLD'S LARGEST MUSLIM POPULATIONS
2009 estimate in millions

Country	Population (millions)
Indonesia	207
Pakistan	171
India	145
Bangladesh	138
Turkey	76
Egypt	75
Nigeria	75
Iran	65

Source: CIA World Factbook

14. About how many times larger is Indonesia's Muslim population than Iran's?

15. Which country in Southeast Asia probably had the most contact with Arab traders in the 1300s? Explain your answer.

SECTION 2 GOVERNMENT & ECONOMICS

2.1 Governing Fragmented Countries

Use with Southeast Asia Today, Section 2.1, *in your textbook.*

BLACK-AND-WHITE COPY READY

Reading and Note-Taking Outline and Take Notes

As you read Section 2.1, create an outline to help you take notes about Southeast Asia's physically fragmented countries. Provide three of the section's major ideas and at least two details to support each of these ideas.

I. Indonesia's struggle for unity

A. 17,000 islands with more than 300 ethnic groups

B.

II.

A.

B.

III.

A.

B.

Evaluate Based on what you have read, which country seems to have had the most success unifying its population? Explain your answer.

BLACK-AND-WHITE COPY READY

Name Class Date

SECTION 2 GOVERNMENT & ECONOMICS

2.1 Governing Fragmented Countries

Use with Southeast Asia Today, Section 2.1, *in your textbook.*

Vocabulary Practice

KEY VOCABULARY

- **fragmented** (FRAG-mehn-tuhd) **country** n., a country that is physically divided into separate parts
- **motto** (MAH-toh) n., a short saying or expression that guides an individual, an organization, or a country

Definition and Details Complete a Definition and Details Chart for each Key Vocabulary word. Provide the definition for each word and add two relevant details from the section that help explain the concept.

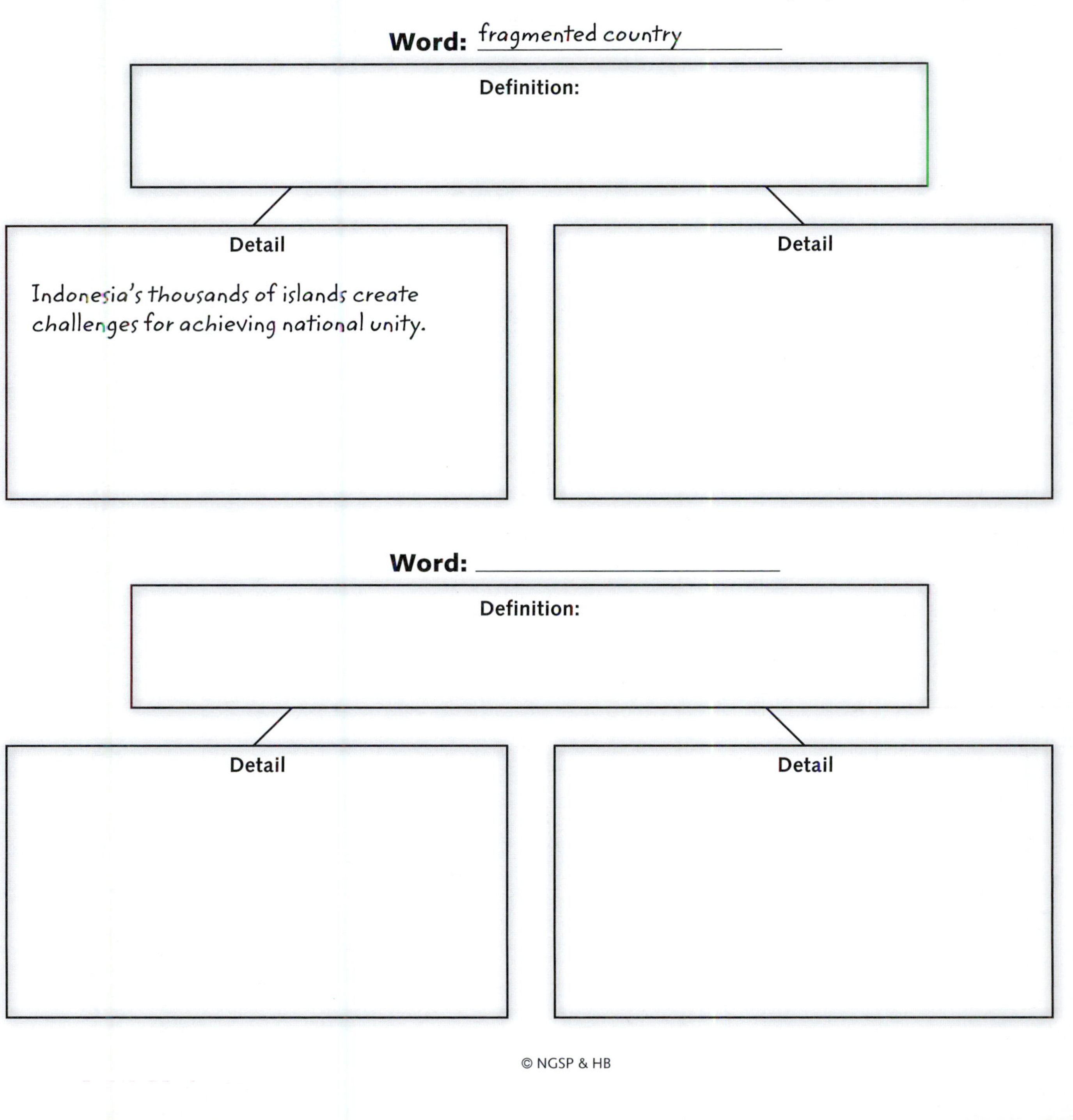

Name Class Date

SECTION 2 GOVERNMENT & ECONOMICS

GeoActivity

Use with Southeast Asia Today, Section 2.1, *in your textbook.*

Go to Interactive Whiteboard GeoActivities *at* **myNGconnect.com** *to complete this activity online.*

NATIONAL GEOGRAPHIC School Publishing

2.1 GOVERNING FRAGMENTED COUNTRIES

Analyze Remittances and GDP

As you have learned, poverty and a lack of jobs have caused many Filipinos to leave their country. In 2009, nearly one tenth of the population was working overseas and sending home remittances. How important are remittances to the Philippines' economy? The chart below shows the GDP and the amount of remittances received over the past 30 years. Study the chart and then answer the questions.

Philippines' GDP and Remittances

YEAR	GDP (billions of U.S. dollars)	REMITTANCES RECEIVED (billions of U.S. dollars)
1979	27.5	0.6
1989	42.6	1.4
1999	76.1	6.7
2009	161.1	19.8

Source: The World Bank

1. **Analyze Data** What happened to GDP and remittances between 1979 and 2009?

2. **Calculate** Remittances are included in a country's GDP. For each year in the chart, remittances made up what percent of GDP? To calculate the percentage, divide the amount of remittances by the GDP and multiply by 100.

3. **Create Graphs** On the grid below, create a bar graph that shows the percentage of GDP made up of remittances. Put the years on the *x*-axis and the amounts of money on the *y*-axis. Decide what the scale should be for the *y*-axis. Be sure to label each axis and give the graph a title.

4. **Interpret Graphs** What trend do you notice about the percentage of GDP made up by remittances between 1979 and 2009?

5. **Draw Conclusions** Is it a good practice for a country to rely on remittances to contribute to the GDP? Why or why not?

Name Class Date

SECTION 2 GOVERNMENT & ECONOMICS

2.2 Migration Within Indonesia

Use with Southeast Asia Today, Section 2.2, *in your textbook.*

BLACK-AND-WHITE COPY READY

Reading and Note-Taking Summarize Information

Use the charts below as you read Section 2.2 to take notes about internal migration in Indonesia.

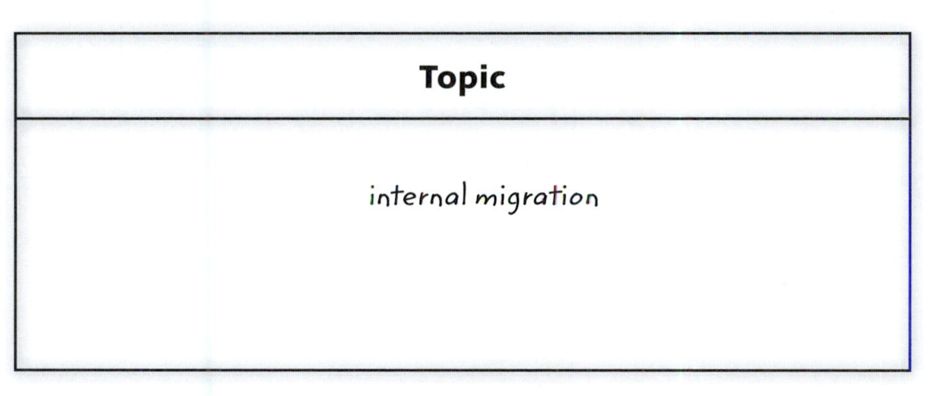

Topic
internal migration

What	Why
The Dutch began relocating people from the inner islands to the outer islands.	

Analyze Cause and Effect How do economic differences between Java and the outer islands affect population patterns in Indonesia?

____

____

Name Class Date

SECTION 2 GOVERNMENT & ECONOMICS

2.2 Migration Within Indonesia

Use with Southeast Asia Today, Section 2.2, *in your textbook.*

Vocabulary Practice

KEY VOCABULARY

- **relocate** (ree-LOH-kayt) v., to move someone or something to a new location
- **trend** n., a change over time in a general direction

Words in Context Follow the directions below for using the Key Vocabulary in writing.

1. Write a sentence using the word *relocate*.

2. Describe which group or groups the Indonesian government wanted to *relocate* through the internal migration program.

3. Identify which groups have chosen to *relocate* despite the goals of the government's internal migration program.

4. Write a sentence using the word *trend*.

5. What *trend* has affected population density on Java and Bali?

6. Write a sentence using both Key Vocabulary words.

Name Class Date

SECTION 2 GOVERNMENT & ECONOMICS

GeoActivity

Use with Southeast Asia Today, Section 2.2, *in your textbook.*

Go to Interactive Whiteboard GeoActivities *at* **myNGconnect.com** *to complete this activity online.*

NATIONAL GEOGRAPHIC School Publishing

2.2 MIGRATION WITHIN INDONESIA

Evaluate Internal Migration

Indonesia's relocation policy has had mixed results and has led to a number of unintended consequences. How has this policy affected the lives of the different groups involved? Review Section 2.2 in your textbook before completing the chart below. Show one of the policy's benefits and drawbacks for each segment of Indonesian society. Then answer the questions.

	BENEFITS	DRAWBACKS
New Settlers	More living space on the outer islands	
Native Population		
Populations of Java and Bali		
Indonesian Government		

1. **Categorize** What is the strongest push factor causing people to move from Indonesia's inner islands to the outer islands? What is the strongest pull factor?

2. **Synthesize** Large sections of Indonesia's hardwood rain forests were cleared to create farmland for settlers on the outer islands. What problems might this deforestation have caused?

3. **Draw Conclusions** Many people from the rural outer islands are moving into Indonesia's already crowded cities on the inner islands in search of work. How must standards of living in these cities compare with standards of living in the rural areas?

2.3 Singapore's Growth

Use with Southeast Asia Today, Section 2.3, *in your textbook.*

Reading and Note-Taking Analyze Cause and Effect

Use the Cause-and-Effect Chain below as you read Section 2.3 to take notes about Singapore's economic growth over the past several decades.

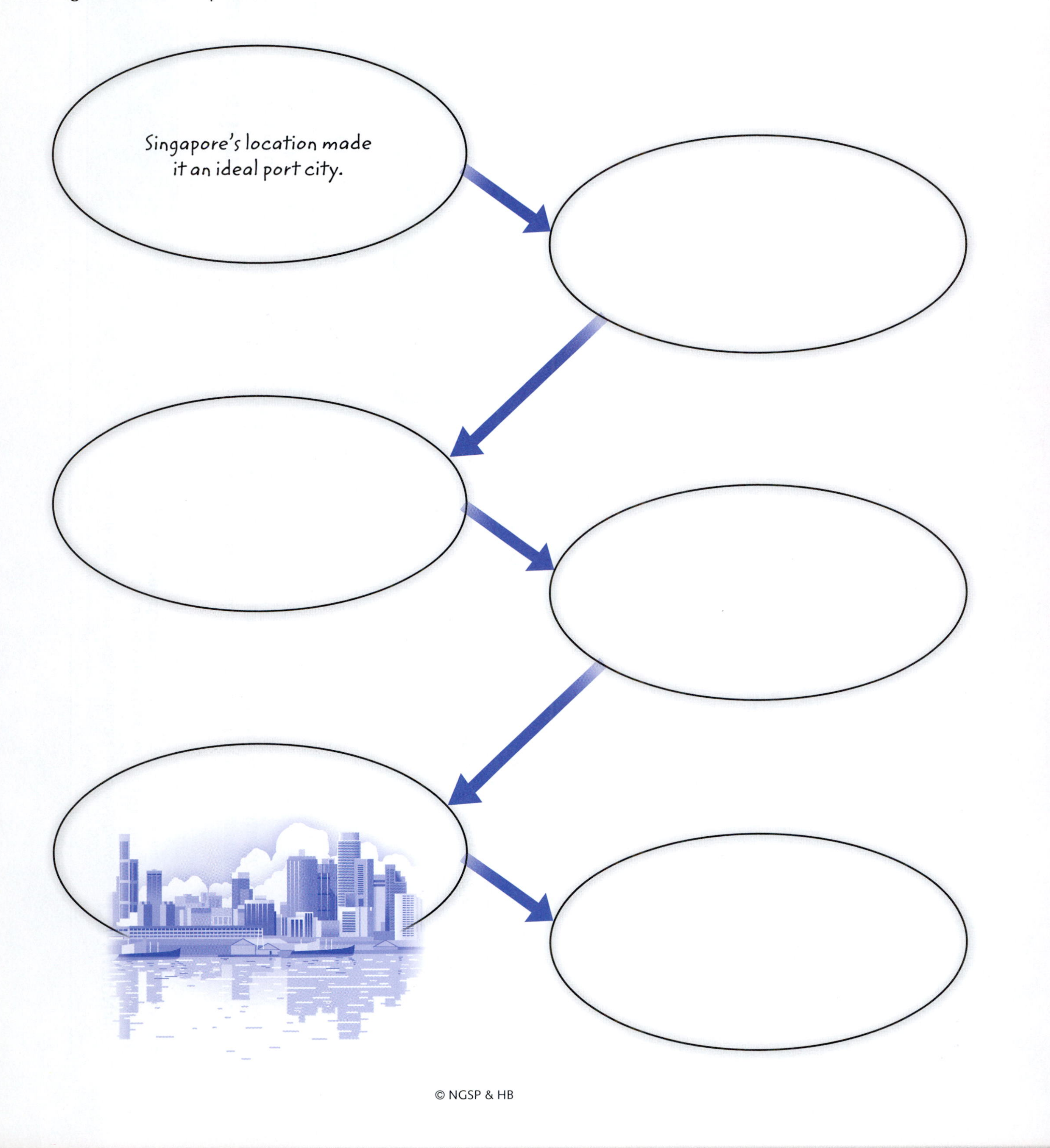

Name Class Date

SECTION 2 GOVERNMENT & ECONOMICS

2.3 Singapore's Growth

Use with Southeast Asia Today, Section 2.3, *in your textbook.*

Vocabulary Practice

KEY VOCABULARY

- **industrialize** (ihn-DUHS-tree-uh-lyz) v., to develop manufacturing on a wide scale
- **multinational corporation** (muhl-tih-NAHSH-nuhl KAWR-puh-ray-shuhn) n., a large business that has operations in many different countries
- **port** n., a town or city with a harbor where ships can exchange cargo

ACADEMIC VOCABULARY

- **potential** (puh-TEHN-shuhl) n., something that exists as a possibility and that may become actual

Write a sentence using the Academic Vocabulary word *potential*.

____

____

Related Idea Web Inside each circle, write one of the Key Vocabulary words and its definition. On the lines connecting the circles, write a sentence about how the two words are related.

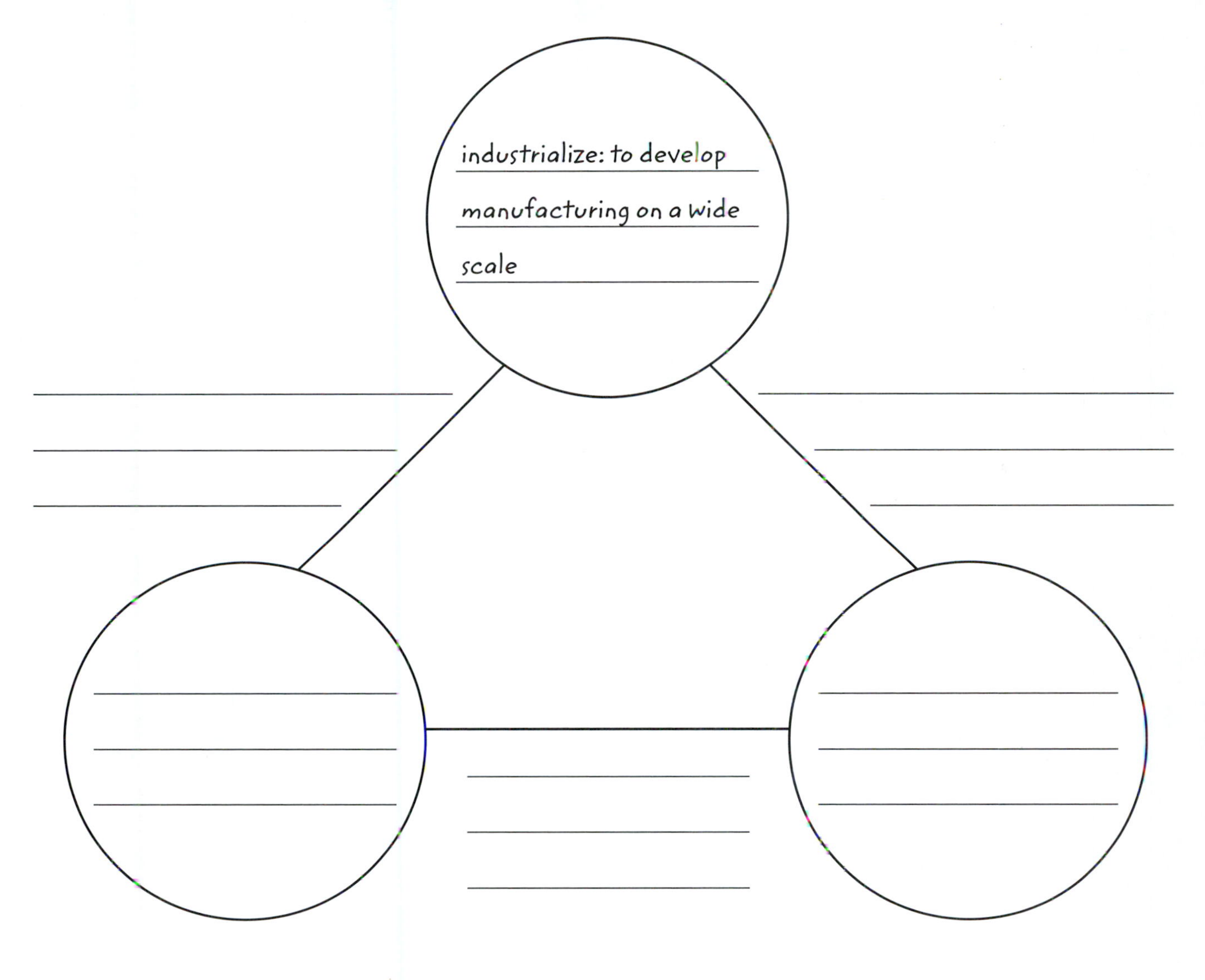

Name ________________ Class ________________ Date ________________

SECTION 2 GOVERNMENT & ECONOMICS

GeoActivity

Use with Southeast Asia Today, Section 2.3, *in your textbook.*

Go to Interactive Whiteboard GeoActivities *at* **myNGconnect.com** *to complete this activity online.*

NATIONAL GEOGRAPHIC School Publishing

2.3 GLOBAL ISSUES: SINGAPORE'S GROWTH

Graph Singapore's Economic Rise

As countries across the region have industrialized, Singapore has faced increasing competition for manufacturing jobs. To remain economically competitive, the country hopes to develop a more educated workforce that will be successful in the areas of technology and finance. To do this, the government has focused heavily on education. How have Singapore's standards of living changed as education has improved? Use the data below to create a line graph at right that shows changes to GDP per capita and university enrollment over the past several decades. Then answer the questions.

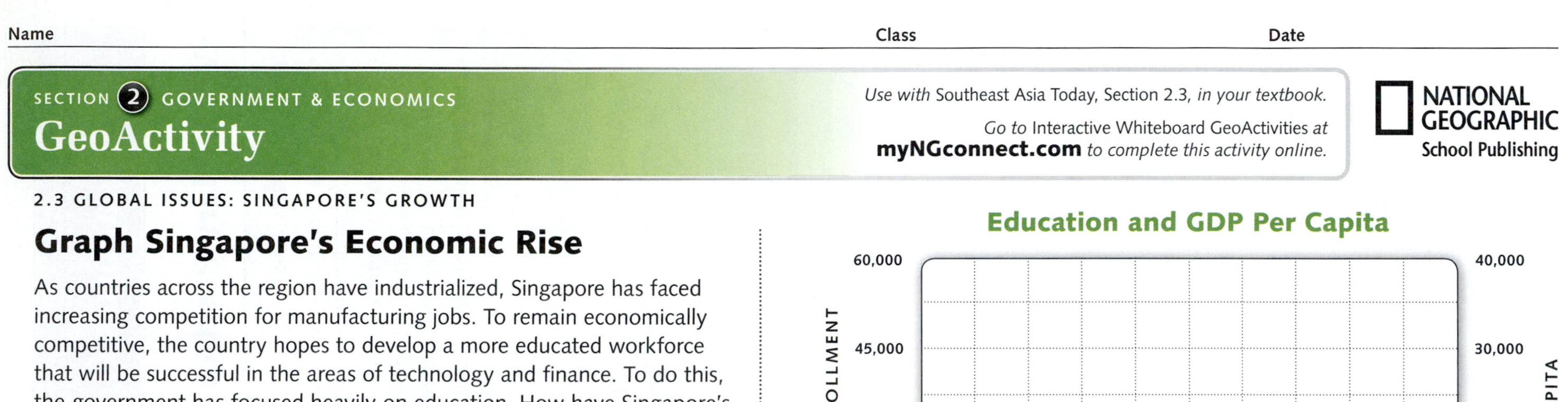

YEAR	UNIVERSITY ENROLLMENT
1980	8,600
1990	22,000
2000	36,100
2009	53,600

Source: Singapore Ministry of Education

YEAR	GDP PER CAPITA (CURRENT U.S. DOLLARS)
1980	4,900
1990	12,100
2000	23,000
2009	36,500

Source: The World Bank

1. **Create Graphs** Use the data from the charts above to graph the university enrollment and GDP per capita between 1980 and 2009 at right. Use the left y-axis for university enrollment and the right y-axis for GDP per capita.

Education and GDP Per Capita

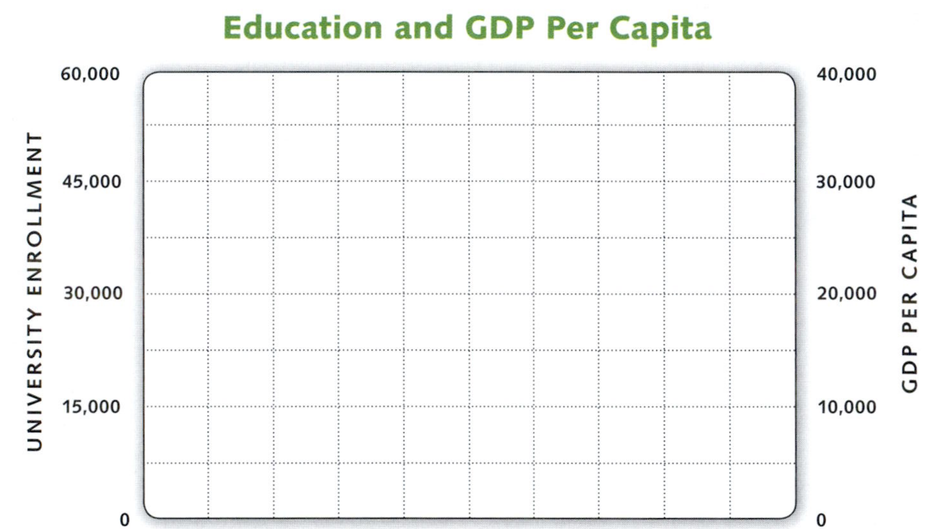

2. **Make Calculations** By 2009, Singapore's population was twice the size it was in 1980, increasing from 2.4 million to 4.7 million people. In that same time, by what amount did university enrollment increase? How does this compare to the population growth?

____

____

3. **Draw Conclusions** Based on your graph, there has been a clear relationship between the two trends. Why might they be related? Explain your reasoning.

____

____

Name Class Date

SECTION 2 GOVERNMENT & ECONOMICS

2.4 Malaysia and New Media

Use with Southeast Asia Today, Section 2.4, *in your textbook.*

Reading and Note-Taking Prediction Map

Use the Prediction Map below as you read Section 2.4 about press freedoms in Malaysia. Then make a prediction about how conditions in Malaysia may change in the decade to come.

Topic:
New Media in Malaysia

Event 1:

Event 2:

Event 3:

Prediction:

2.4 Malaysia and New Media

Use with Southeast Asia Today, Section 2.4, *in your textbook.*

Vocabulary Practice

KEY VOCABULARY

- **emergence** (ih-MUHR-jihnts) n., the process of coming into being or arriving
- **reliable** (rih-LY-uh-buhl) adj., trustworthy or dependable

WDS Triangles Complete a WDS Triangle for the Key Vocabulary words *emergence* and *reliable*. Beside the "W," write the word. Beside the "D," write the definition. Beside the "S," write a sentence that uses the Key Vocabulary word.

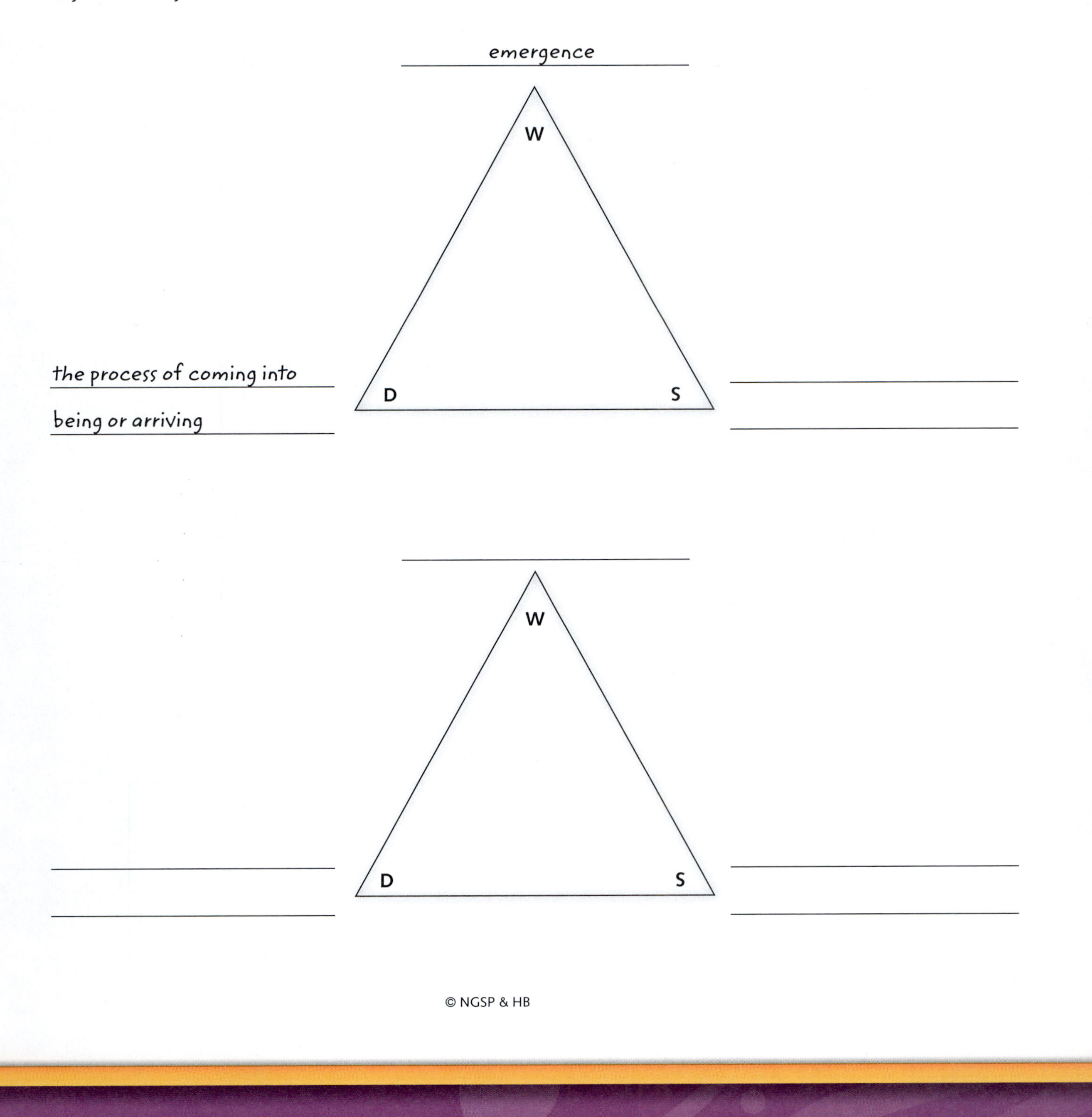

Name _______________ Class _______________ Date _______________

SECTION 2 GOVERNMENT & ECONOMICS

GeoActivity

Use with Southeast Asia Today, Section 2.4, *in your textbook.*
Go to Interactive Whiteboard GeoActivities *at* **myNGconnect.com** *to complete this activity online.*

NATIONAL GEOGRAPHIC School Publishing

2.4 MALAYSIA AND NEW MEDIA

Explore Effects of New Media

New media continues to change the way people communicate. It has also influenced the relationship between governments and their citizens. Read about three political situations where new media played a leading role. Then answer the questions.

Malaysia, 2008 Malaysia's government keeps tight control over newspapers, radio stations, and television stations. However, in the 1990s, the government agreed not to control the Internet, hoping this policy would attract international investors. This Internet freedom allowed people to speak out against the government on Malaysiakini and other Web sites. These sites attracted more readers than the tightly-controlled newspapers. They are believed to have influenced the 2008 national election, when the long-ruling United Malay National Organization suffered its worst defeat since the country's independence in 1957.

1. **Analyze Cause and Effect** What effect has the tightly-controlled media had on Malaysian politics? How did the Internet change this?

____

____

____

United States, 2008 In the U.S. presidential election of 2008, the Internet played a much larger role than it ever had before. Campaign Web sites and Web videos allowed candidates to share their ideas directly with their supporters rather than relying on traditional media coverage. These Web sites also allowed candidates to conduct online fundraising and to organize events. Political blogs became popular and allowed people outside of either the campaigns or traditional media to share ideas and debate issues online.

2. **Evaluate** Why might political candidates and voters continue to rely in part on traditional media if they are able to bypass the mainstream media and communicate directly with their supporters?

____

____

____

____

Egypt, 2011 Different forms of social media were important tools in the 2011 revolution in Egypt. After Egypt's government realized how the Internet and social networking sites were helping Egyptians communicate about the revolution, it shut down the Internet and cell phone coverage. Only about 20 percent of Egyptians have access to the Internet, but that access was enough to overpower the embattled government. Once Internet access was restored, Egyptians shared real-time videos, updates, and observations about the remarkable events that toppled Hosni Mubarak's 30-year regime.

3. **Make Predictions** How will the continued development of digital media, social networking, and mobile technology affect the ability of governments to control information?

____

____

____

____

Name Class Date

SECTION 2 GOVERNMENT & ECONOMICS

2.1–2.4 Review and Assessment

Use with Southeast Asia Today, Sections 2.1–2.4, *in your textbook.*

BLACK-AND-WHITE COPY READY

Follow the instructions below to review what you have learned in this section.

Vocabulary Next to each vocabulary word, write the letter of the correct definition.

1.____ fragmented country
2.____ motto
3.____ relocate
4.____ port
5.____ industrialize
6.____ multinational corporation
7.____ emergence
8.____ reliable

A. to move someone or something to a new location
B. to develop manufacturing on a wide scale
C. a country that is physically divided into separate parts
D. trustworthy or dependable
E. a town or city with a harbor where ships can exchange cargo
F. a short saying or expression that guides an individual, an organization, or a country
G. the process of coming into being or arriving
H. a large business that has operations in many different countries

Main Ideas Use what you've learned about Southeast Asia's governments and economies to answer these questions.

9. Compare and Contrast In what ways are the islands of Java and Sumatra different?

____

10. Identify Which two areas of land make up Malaysia?

____

11. Analyze Cause and Effect How has the widespread use of the Tagalog language affected Filipino society?

____

12. Categorize Are Java, Madura, and Bali known as the inner islands or the outer islands?

____

13. Sequence Events Did internal migration begin before or after the Indonesian people gained control of the country from the Dutch?

____

14. Evaluate What has been the result of Indonesia's internal migration program?

____

15. Explain Why has crowding in Java and Bali grown worse in recent years?

____

16. Identify Before gaining its independence, Singapore was a part of what country?

____

17. Compare and Contrast How did Singapore's economic output per person compare with Malaysia's in 2003?

____

18. Analyze Cause and Effect How did the transfer of control of Hong Kong from Britain to China affect Singapore economically?

____

Focus Skill: Make Generalizations

Answer the questions below to make generalizations about Southeast Asia and the ways the region's governments try to unify their countries.

19. How does the physical geography of Indonesia, Malaysia, and the Philippines affect national unity?

20. How does economic development affect national unity?

21. Why can a diversity of ethnic groups threaten national unity in a country?

22. How does Singapore's location benefit its economy?

23. Why is an educated workforce important to an industrialized country?

24. What does Singapore's government do to attract foreign investment?

25. Using what you have learned about Singapore and India, how do you think education levels relate to standards of living?

26. How do government restrictions on Malaysian newspapers affect the type of information they publish?

Synthesize: Answer the Essential Question

How are Southeast Asia's governments trying to unify their countries? Think about the challenges that the region's fragmented countries face. Recall what you have read about efforts to unify these countries economically and culturally.

SECTION 2 GOVERNMENT & ECONOMICS

2.1–2.4 Standardized Test Practice

Use with Southeast Asia Today, Sections 2.1–2.4, *in your textbook.*

Follow the instructions below to practice test-taking on what you've learned from this section.

Multiple Choice Circle the best answer for each question from the options available.

1. About how many languages are spoken in Indonesia?

A 25
B 125
C 250
D 1500

2. "Diversity in unity" is the motto of which country?

A the Philippines
B Indonesia
C Malaysia
D India

3. Malaysia has large minority populations from which two countries?

A Japan and South Korea
B Pakistan and Bangladesh
C Australia and New Zealand
D China and India

4. How has the strong progress toward becoming a developed nation affected ethnic tensions in Malaysia?

A The economic growth has helped ease tensions.
B The economic growth has led to increased competition and resentment.
C The economic growth has drawn more immigrants to the country, increasing ethnic hostilities.
D The economic growth has not affected relationships among ethnic groups.

5. Where does Indonesia rank among countries with the largest populations in the world?

A 1st
B 4th
C 22nd
D 49th

6. Moving Javanese people to other islands would meet what goal of the Indonesian government?

A to create businesses and bring prosperity to remote areas
B to create new majorities for the ruling political party in these areas
C to move young people out of the cities
D to spread the official language and help unify the country

7. Why did the British establish the port of Singapore?

A to establish a military base in the region
B to spread Christianity
C to compete with the Dutch in trade
D to learn about local cultures

8. Singapore's leaders have set goals based largely on what?

A economic success
B cultural greatness
C social equality
D political freedoms

9. Since gaining independence in 1963, Malaysian citizens have enjoyed which of the following?

A high economic growth and high levels of political freedom
B high economic growth but limited political freedoms
C low economic growth but high levels of political freedom
D low economic growth and limited political freedoms

10. What media type has given Malaysians access to more accurate information in recent years?

A newspaper
B radio station
C television station
D web site

Document-Based Question

Read the passage on Singapore from your *World Cultures and Geography* textbook. Then answer the questions below.

> *Singapore thrived because of its prime location. It served as the main transit point for sending raw materials such as timber, rubber, rice, and petroleum from Southeast Asia to other parts of the world. Manufactured goods from the United States and Europe came into the port and were shipped to other ports in Southeast Asia. Cars and machinery were shipped into the city from the west to be distributed around the region.*

Source: *National Geographic World Cultures and Geography*, 2013

Constructed Response

11. According to the passage, why did Singapore thrive as a port?

____

12. What kinds of materials were shipped to and from Singapore?

____

Extended Response

13. Explain how Singapore is part of the global economy.

____

____

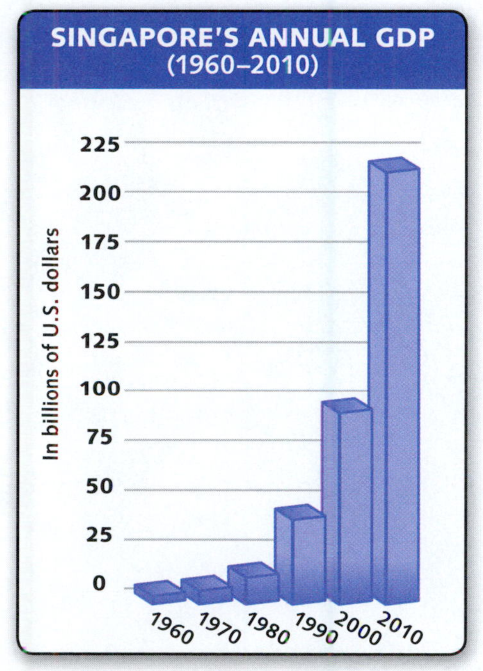

Source: CIA World Factbook

14. By how much did Singapore's GDP increase between 2000 and 2010?

____

____

15. Based on the graph and what you have read, have Singapore's policies designed to strengthen its economy been successful? How could other countries in the region achieve results similar to these?

____

____

Southeast Asia

Southeast Asia Geography & History

QUIZ: SECTION 1 GEOGRAPHY

Multiple Choice Circle the best answer for each question from the choices available.

1 What country sits entirely inland, with no coasts?
- **A** Cambodia
- **B** Laos
- **C** Myanmar
- **D** Vietnam

2 Which natural event can take place because of the wet monsoon season?
- **A** earthquake
- **B** tsunami
- **C** typhoon
- **D** volcano

3 The Mekong River, like other rivers, creates what landform as it approaches the sea?
- **A** delta
- **B** harbor
- **C** plain
- **D** valley

4 In 1989, what made the government of Malaysia put a ban on the harvesting of teak wood?
- **A** a growing number of tourists wanting to visit the forests
- **B** a drop in the value of the product
- **C** a growing concern about the animals living in the forests
- **D** a landslide caused by deforestation

5 In Indonesia and the Philippines, where do most people live?
- **A** by inland forests
- **B** by large rivers
- **C** on steep mountains
- **D** on lowland plains

6 In what area are scientists discovering new species?
- **A** Foja Mountains
- **B** Malay Peninsula
- **C** Mekong Delta
- **D** Spice Islands

Constructed Response Write the answer to each question in the space provided.

7 What geographic event 6,000 years ago caused land bridges in Southeast Asia to disappear?

____

____

8 What accounts for the many geographic changes that still occur in the area?

____

____

Name Class Date

Southeast Asia Geography & History

QUIZ: SECTION 2 HISTORY

Multiple Choice Circle the best answer for each question from the choices available.

1 What two bodies of water are connected by waterways in Southeast Asia?
- A South China and Arafura Seas
- B Pacific and Indian Oceans
- C Philippine and Andaman Seas
- D Atlantic and Pacific Oceans

2 At about the same time that Chinese culture first came to Southeast Asia, what other culture had influences in the region?
- A Australian
- B Dutch
- C Indian
- D Japanese

3 In the 1500s, what product attracted Europeans to the region?
- A bananas
- B bauxite
- C spices
- D wood

4 Fossils found on what island suggest that humans lived there more than one million years ago?
- A Borneo
- B Java
- C Sumatra
- D Timor

5 For what country did Emilio Aguinaldo want to gain independence?
- A Indonesia
- B Malaysia
- C Philippines
- D Vietnam

6 In what Southeast Asian country did the United States fight a war from 1964 to 1973?
- A Cambodia
- B Indonesia
- C Thailand
- D Vietnam

Constructed Response Write the answer to each question in the space provided.

7 Centuries before European colonialism, the influence of nearby civilizations caused the spread of what two religions across Southeast Asia?

____

____

8 What influence has trade had on the history of the region?

____

____

Name Class Date

Southeast Asia Geography & History

CHAPTER TEST A

Part 1: Multiple Choice Circle the best answer for each question from the choices available.

1 What factor makes rice a successful crop in Thailand's Central Plain?
- **A** Villages provide the region with the workers needed for rice farming.
- **B** Sediment deposits from rivers create fertile soil for rice farming.
- **C** Inland breezes from nearby seas create ideal temperatures for rice crops.
- **D** Typhoons drench the region every year, flooding rice crops.

2 What city in Thailand sits along the banks of the Chao Phraya River?
- **A** Bangkok
- **B** Ho Chi Minh City
- **C** Jakarta
- **D** Phnom Penh

3 Which of the following problems does Malaysia face today?
- **A** deforestation
- **B** desalinization
- **C** dictatorship
- **D** unemployment

4 Which of the following statements is true about island countries in the region?
- **A** They have limited exports to trade with other countries.
- **B** They are sparsely populated.
- **C** They are located where tectonic plates come together.
- **D** They sit at or below sea level.

5 What makes the Foja Mountains of New Guinea a unique place for scientists to do research?
- **A** Plant species are similar to those found in northern climates.
- **B** Animal species have lived there without human presence.
- **C** Animal species have increased in number over the past decade.
- **D** Plant species are known to survive volcanic activity.

6 What was the largest and longest lasting mainland empire in ancient Southeast Asia?
- **A** Dai Viet Empire of Vietnam
- **B** Khmer Empire of Cambodia
- **C** Majapahit Empire of Java
- **D** Srivijaya Empire of Sumatra

7 In the 17th century, the Netherlands gained power in which country in the region?
- **A** Indonesia
- **B** Philippines
- **C** Thailand
- **D** Vietnam

8 During World War II, what country seized control of both Indonesia and the Philippines?
- **A** Germany
- **B** Great Britain
- **C** Japan
- **D** United States

9 When did the Philippines gain its independence?
- **A** 1571
- **B** 1898
- **C** 1946
- **D** 1974

10 Why did the United States get involved in the Vietnam War?
- **A** It feared that North Vietnam would spread communism in the region.
- **B** It wanted to maintain good relations with South Vietnam.
- **C** It wanted to protect important trade routes from being cut off.
- **D** It feared that the ruling government would cause poverty in the region.

BLACK-AND-WHITE COPY READY

BLACK-AND-WHITE COPY READY

Name ______________________ Class ______________ Date ______________

Southeast Asia Geography & History

CHAPTER TEST A

Part 2: Interpret Maps Use the map and your knowledge of Southeast Asia to answer the questions below.

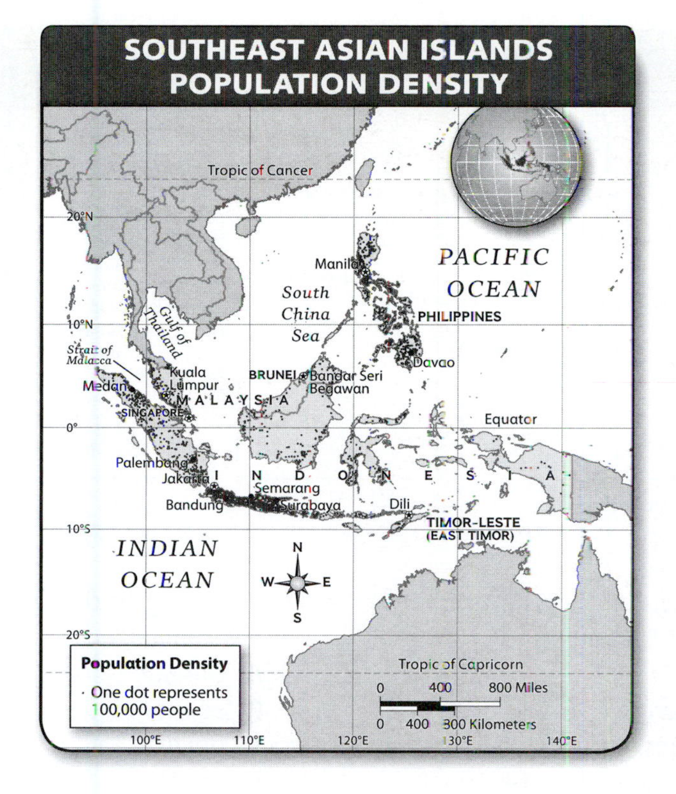

11 Which of the following cities is located on the most densely populated island?

- **A** Bandung
- **B** Davao
- **C** Manila
- **D** Singapore

12 What location factor do the densely populated cities of Medan, Jakarta, and Manila have in common?

- **A** They are near the Malay Peninsula.
- **B** They are on a coast.
- **C** They are near the South China Sea.
- **D** They are on a river.

13 Based on the map, which of the following statements is true?

- **A** Population density is greater on Pacific Ocean islands than on Indian Ocean Islands.
- **B** Population density is greatest on islands 10° north of the equator.
- **C** Population density is less at the equator than it is 10° north and south of the equator.
- **D** Population density is evenly distributed across the region.

Constructed Response Use a complete sentence to write the answer in the space provided.

14 How would you describe the location of Indonesia's most populated island?

____

____

Name Class Date

Southeast Asia Geography & History

CHAPTER TEST A

Part 3: Interpret Charts Use the chart and your knowledge of Southeast Asia to answer the questions below.

THE PLANTS, ANIMALS, AND PEOPLE OF BORNEO AND SUMATRA

Geography	Selected Animals	Plants	People	Challenges
• Islands • On the equator • Tropical climate • Diverse rain forests • Diverse eco regions	• Tigers • Orangutans • Pygmy elephants • Rhinos • Proboscis monkeys • Sun bears • Clouded leopards • Flying fox bat	• Over 15,000 known plant species • More than 400 plant species discovered since 1995	Combined population of more than 56 million includes both immigrants and indigenous people. Native groups have lived off the forests for generations.	• Commercial logging • Conversion of forest land to agricultural land

Source: World Wildlife Federation

15 Based on the chart, which of the following statements is true about Borneo's and Sumatra's populations?
A Young people are leaving home to find work in mainland cities.
B New immigrants are moving to the rain forests.
C Indigenous populations have lived off forests for a long time.
D Island populations have increased rapidly.

16 Which of the following factors supports the idea that Borneo and Sumatra have a tropical climate?
A They are located south of the mainland.
B They include many diverse ecoregions.
C They have a variety of animal species.
D They are located exactly on the equator.

17 According to the chart, what type of change has challenged traditional life in recent years?
A climatic
B economic
C political
D social

Constructed Response Use a complete sentence to write the answer in the space provided.

18 What generalization can be made about Borneo's and Sumatra's natural environments?

____

____

BLACK-AND-WHITE COPY READY

BLACK-AND-WHITE COPY READY

Name Class Date

Southeast Asia Geography & History

CHAPTER TEST A

Part 4: Document-Based Question Use the documents and your knowledge of Southeast Asia to answer the questions below.

Introduction

Southeast Asia's geography includes natural habitats that harbor many unique plants and animals, with some settings so remote that previously unknown species have recently been discovered. The country's landforms also include an abundance of natural resources. The desire for economic growth and development sometimes threatens the unique habitats in which wild plants and animals are found.

Objective: Analyze the relationship between preserving the rain forest and economic growth.

DOCUMENT 1 Passage describing rain forest resources

Some of the trees native to the Southeast Asian rain forest have significant economic value, and as a result large areas of rain forest have been cleared to plant them. At one time, teak wood from Thailand was a large part of that country's economy. However, after a landslide in 1989, which was blamed on excessive deforestation, the government imposed a ban on harvesting teak.

Source: *National Geographic World Cultures and Geography*, 2013

Constructed Response Use complete sentences to write the answers in the space provided.

19 What was the reason the rain forest was cleared?

____

____

20 What effect did clearing the rain forest have on some of the land in Thailand?

____

____

DOCUMENT 2 Visual of Malaysia's currency

Source: *Complete National Geographic*, October 2006

Constructed Response Use complete sentences to write the answers in the space provided.

21 Based on the symbols chosen for the Malaysian currency, what conclusions can be reached about the government's view of what the country's future should be?

---

22 Based on the information given in Documents 1 and 2, what inference might be made about the attitude the region's governments have toward economic development?

DOCUMENT 3 Passage describing Malaysia's biosphere reserve [protected ecosystem]

Tasik Chini, Malaysia, is the country's first site as a UNESCO designated [official] biosphere reserve. The lake [in the center of the reserve] is a sanctuary [safe place] for many endemic [native] freshwater species on which intensive research and monitoring is carried out by various research institutions. Handicraft production (such as textiles) around the lake and its tributary rivers are seen to have great development potential for the larger area.

Source: David Braun, "United Nations Proclaims 22 New Biosphere Reserves," *National Geographic News Watch*, May 27, 2009 http://blogs.nationalgeographic.com/blogs/news/chiefeditor/2009/05/united-nations-proclaims-22-ne.html#comments

Constructed Response Use complete sentences to write the answers in the space provided.

23 Why is the Tasik Chini Lake considered an important place?

---

24 What additional information in the passage reveals a potential conflict with the efforts of researchers?

Extended Response Write a paragraph to answer the question. Use information from all three documents and your knowledge of Southeast Asia in writing your paragraph. Use the back of this page or a separate piece of paper to write your answer.

25 In your opinion, is it possible for countries in the region to strike a balance between the desire for economic growth and the need to preserve the environment?

BLACK-AND-WHITE COPY READY

Name Class Date

Southeast Asia Geography & History

CHAPTER TEST B

Part 1: Multiple Choice Circle the best answer for each question from the choices available.

1 Which statement best describes the region's climate?
- A Its climate includes both arid and humid conditions.
- B It has a moderate climate.
- C Its climate includes both cold and hot extremes.
- D It has a tropical climate.

2 What term is used when people fish for only what they need to live?
- A commercial fishing
- B ecological fishing
- C seasonal fishing
- D subsistence fishing

3 What geographic feature connects the Indian and Pacific Oceans on either side of the Malay Peninsula?
- A delta
- B reef
- C river
- D strait

4 Which of these statements about the region's island volcanoes is true?
- A Many island volcanoes have been inactive for a long time.
- B Most island volcanoes are still active.
- C Erupting volcanoes often create poor soil conditions.
- D People avoid living near volcanoes.

5 What have scientists discovered in the Foja Mountains of New Guinea?
- A rare mineral and oil reserves
- B unusual climate patterns
- C new plant and animal species
- D prehistoric fossil remains

6 What helped certain kingdoms In ancient Southeast Asia gain power?
- A control of trade
- B farming success
- C spread of religion
- D language teaching

7 Which of the following countries played a role in colonizing Southeast Asia?
- A Belgium
- B France
- C Germany
- D Russia

8 Part of what country was known to Europeans as the Spice Islands?
- A Cambodia
- B Indonesia
- C Malaysia
- D Philippines

9 What city has remained the economic, political, and cultural center of the Philippines for centuries?
- A Bangkok
- B Davao
- C Jakarta
- D Manila

10 In 1959, why did North Vietnam's president send aid to overthrow South Vietnam's government?
- A He hoped to create a single communist country.
- B He learned of a plan to overthrow his regime.
- C He hoped to improve the economy of the area.
- D He learned of a plan for further colonization.

Southeast Asia Geography & History

CHAPTER TEST B

Part 2: Interpret Maps Use the map and your knowledge of Southeast Asia to answer the questions below.

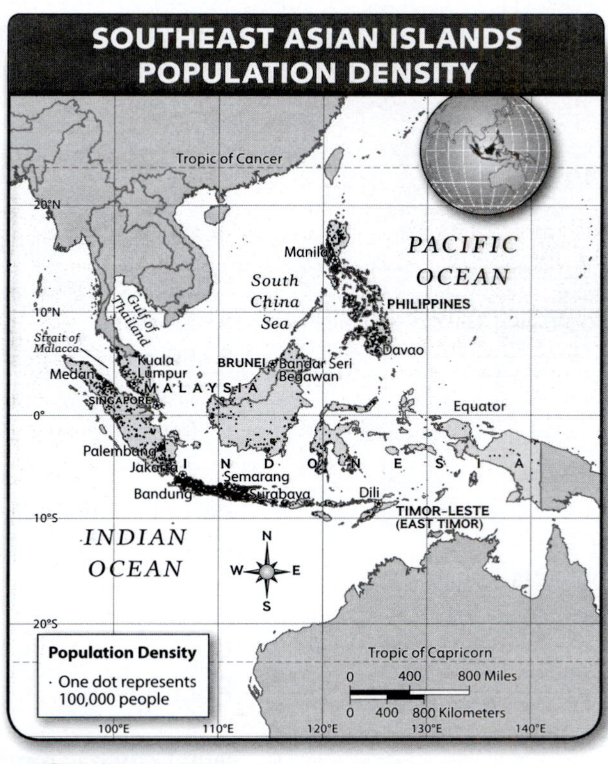

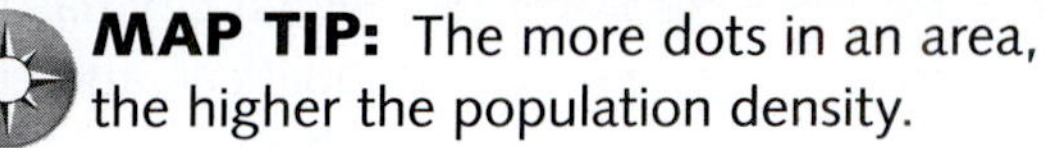
MAP TIP: The more dots in an area, the higher the population density.

11 On the map, what densely populated city is the farthest north?
- **A** Davao
- **B** Jakarta
- **C** Manila
- **D** Medan

12 Which area has the lowest population density?
- **A** western Malaysia
- **B** southern Indonesia
- **C** northern Philippines
- **D** eastern Indonesia

13 The densely populated cities of Davao, Semarang, and Jakarta have what in common?
- **A** They are near the Indian Ocean.
- **B** They are in coastal areas.
- **C** They are near the Pacific Ocean.
- **D** They are on the same island.

14 What densely populated city is located closest to the equator?
- **A** Davao
- **B** Dili
- **C** Manila
- **D** Medan

Name Class Date

Southeast Asia Geography & History

CHAPTER TEST B

Part 3: Interpret Charts Use the chart and your knowledge of Southeast Asia to answer the questions below.

THE PLANTS, ANIMALS, AND PEOPLE OF BORNEO AND SUMATRA

Geography	Selected Animals	Plants	People	Challenges
• Islands • On the equator • Tropical climate • Diverse rain forests • Diverse eco regions	• Tigers • Orangutans • Pygmy elephants • Rhinos • Proboscis monkeys • Sun bears • Clouded leopards • Flying fox bat	• Over 15,000 known plant species • More than 400 plant species discovered since 1995	Combined population of more than 56 million includes both immigrants and indigenous people. Native groups have lived off the forests for generations.	• Commercial logging • Conversion of forest land to agricultural land

Source: World Wildlife Federation

15 Since 1995, how many new plant species have been discovered on Borneo and Sumatra?
- **A** 200
- **B** 400
- **C** 1,000
- **D** 15,000

16 Which of the following animals is found on the two islands?
- **A** clouded leopard
- **B** brown bear
- **C** squirrel monkey
- **D** spotted owl

17 What two words best describe the climate of the islands?
- **A** cold and dry
- **B** cold and humid
- **C** hot and humid
- **D** hot and dry

18 According to the chart, what challenge do the islands face?
- **A** Forests are being destroyed.
- **B** Traditional peoples are leaving.
- **C** Land is scarce for animal life.
- **D** Farms are being taken over.

Southeast Asia Geography & History

CHAPTER TEST B

Part 4: Document-Based Question Use the documents and your knowledge of Southeast Asia to answer the questions below.

Introduction

Southeast Asia's geography includes natural habitats that harbor many unique plants and animals, with some settings so remote that previously unknown species have recently been discovered. The country's landforms also include an abundance of natural resources. The desire for economic growth and development sometimes threatens the unique habitats in which wild plants and animals are found.

Objective: Examine the relationship between preserving the rain forest and economic growth.

DOCUMENT 1 Passage describing rain forest resources

> Some of the trees native to the Southeast Asian rain forest have significant economic value, and as a result large areas of rain forest have been cleared to plant them. At one time, teak wood from Thailand was a large part of that country's economy. However, after a landslide in 1989, which was blamed on excessive deforestation, the government imposed a ban on harvesting teak.
>
> **Source:** *National Geographic World Cultures and Geography*, 2013

Constructed Response Write the answer to each question in the space provided. You do not need to write complete sentences.

19 How did teak wood provide an economic benefit?

____

____

20 What damage resulted from clearing the rain forest?

____

____

DOCUMENT 2 Visual of Malaysia's currency

Source: *Complete National Geographic*, October 2006

Constructed Response Write the answer to each question in the space provided.

21 What symbols are used on the Malaysian currency to stand for transportation?

____

22 What do these symbols show about the government's hopes for the future?

____

DOCUMENT 3 Passage describing Malaysia's biosphere reserve [protected ecosystem]

Tasik Chini, Malaysia, is the country's first site as a UNESCO designated [official] biosphere reserve. The lake [in the center of the reserve] is a sanctuary [safe place] for many endemic [native] freshwater species on which intensive research and monitoring is carried out by various research institutions. Handicraft production (such as textiles) around the lake and its tributary rivers are seen to have great development potential for the larger area.

Source: David Braun, "United Nations Proclaims 22 New Biosphere Reserves," *National Geographic News Watch*, May 27, 2009 http://blogs.nationalgeographic.com/blogs/news/chiefeditor/2009/05/united-nations-proclaims-22-ne.html#comments

Constructed Response Write the answer to each question in the space provided.

23 Why are researchers interested in the Tasik Chini Lake?

____

24 Why do some think this area has great development potential?

____

Extended Response Write a paragraph to answer the questions. Use information from all three documents and your knowledge of South Asia in writing your paragraph. Use the back of this page or a separate piece of paper to write your answer.

25 In your opinion, is it important for countries in the region to preserve the natural environment as they try to develop their economy? Explain.

Southeast Asia Today

QUIZ: SECTION 1 CULTURE

Multiple Choice Circle the best answer for each question from the choices available.

1 In animism, how do people worship?
A attend church on Sundays
B go to mosque on Fridays
C perform rituals to please spirits
D give honor to the national ruler

2 What religion did Europeans spread to Southeast Asia?
A Buddhism
B Christianity
C Hinduism
D Islam

3 About how many young people in Thailand work in cities?
A 1 out of 3
B 2 out of 5
C 3 out of 4
D 4 out of 5

4 What force was most responsible for causing European languages to gain influence in Southeast Asia?
A colonialism
B immigration
C religion
D trade

5 When a regional language disappears, what is lost?
A economic opportunity
B educational progress
C national unity
D traditional culture

6 Why do poachers kill Asian elephants?
A for tusks
B for meat
C for hides
D for bones

Constructed Response Write the answer to each question in the space provided.

7 How has the migration of young people to cities affected religious practices in Thailand?

8 Why do elephants and farmers sometimes come into conflict?

Name Class Date

Southeast Asia Today

QUIZ: SECTION 2 GOVERNMENT & ECONOMICS

Multiple Choice Circle the best answer for each question from the choices available.

1 Which country is divided between a mainland area and an island?
A Indonesia
B Malaysia
C the Philippines
D Timor Leste

2 What is one of the two official languages of the Philippines?
A Chinese
B English
C Malay
D Spanish

3 What type of migration did the Indonesian government promote?
A farms to cities
B outer islands to inner islands
C islands to mainland
D inner islands to outer islands

4 What caused Malaysia and Singapore to separate?
A ethnic tensions
B geographic separation
C colonial histories
D economic competition

5 Why did the government of Singapore invest in infrastructure improvements?
A to gain access to current technologies
B to build green sources of energy
C to improve the country's water supply
D to double electrical production

6 What threatens the ability of the Malaysian government to control information?
A growing literacy rates
B foreign troublemakers
C new media technologies
D political propaganda

Constructed Response Write the answer to each question in the space provided.

7 What geographic factor makes it difficult for the governments of Indonesia, Malaysia, and the Philippines to unify their countries?

____

____

8 What type of society does the government of Singapore want its country to have?

____

____

Name Class Date

Southeast Asia Today

CHAPTER TEST A

Part 1: Multiple Choice Circle the best answer for each question from the choices available.

1 Who brought Islam to Southeast Asia?
- **A** Arab traders
- **B** Chinese conquerors
- **C** European colonizers
- **D** Indian pilgrims

2 Why do traditional Thai buildings have steeply slanted roofs?
- **A** to conform to religious designs
- **B** to keep snow from piling too high
- **C** to use solar energy panels
- **D** to shed heavy monsoon rains

3 What is one result of the migration to cities in Thailand?
- **A** People wear Western clothing.
- **B** More women cook meals for their families.
- **C** People lose their village ties.
- **D** More young men become temporary monks.

4 What statement accurately describes Southeast Asian languages?
- **A** Each country has only one official language.
- **B** The region is home to hundreds of languages.
- **C** Colonial languages wiped out native languages.
- **D** Governments have outlawed the speaking of dialects.

5 What problem most threatens the Asian elephant population?
- **A** climate change
- **B** repeal of protective laws
- **C** loss of habitat
- **D** safaris by tourist hunters

6 What approach have both the Indonesian and Malaysian governments taken to try to unify their countries?
- **A** outlawing the speaking of local dialects
- **B** improving overall standards of living
- **C** expelling minority groups from their borders
- **D** encouraging migration to the capital

7 Why have a large number of Filipinos left their country?
- **A** ethnic violence
- **B** high unemployment
- **C** natural disasters
- **D** severe pollution

8 What was an unwanted effect of relocating Indonesians from the inner islands to the outer islands?
- **A** cultural diffusion
- **B** economic growth
- **C** environmental damage
- **D** urban crowding

9 What geographic characteristic helped Singapore succeed economically?
- **A** fertile volcanic soil
- **B** protection given by mountains
- **C** fast-flowing rivers
- **D** location near trade routes

10 What effect has the arrival of the Internet had in Malaysia?
- **A** It has challenged government versions of the truth.
- **B** It has caused an increase in political fundraising.
- **C** It has led to the arrest of several legislators for corruption.
- **D** It has prompted the overthrow of a military dictator.

Name Class Date

Southeast Asia Today

CHAPTER TEST A

Part 2: Interpret Maps Use the map and your knowledge of Southeast Asia to answer the questions below.

11 On which island is the capital located?
- **A** Luzon
- **B** Mindanao
- **C** Mindoro
- **D** Palawan

12 Based on the map, which island's residents probably feel most isolated from the rest of the country?
- **A** Luzon
- **B** Mindanao
- **C** Mindoro
- **D** Palawan

13 What phrase accurately describes the location of the major cities shown on this map?
- **A** on Luzon
- **B** on Mindoro
- **C** near a coast
- **D** near an island center

Constructed Response Use complete sentences to write the answer in the space provided.

14 For years, rebels on the island of Mindanao have wanted independence from the Filipino government. What geographic reasons help explain their feelings?

____

____

Name Class Date

Southeast Asia Today

CHAPTER TEST A

Part 3: Interpret Graphs Use the graph and your knowledge of Southeast Asia to answer the questions below.

15 Which of the outer islands has the greatest population?

A Borneo
B New Guinea
C Sulawesi
D Sumatra

16 Which of the island groups has the smallest population?

A Bangka and Beltung
B Flores, Sumba, and Timor
C Lombok and Sumbawa
D Molucca Islands

17 How does the population of all the outer islands combined compare to the combined population of the inner islands?

A about one-tenth
B about two-thirds
C about equal
D about double

INDONESIA: ISLAND POPULATIONS

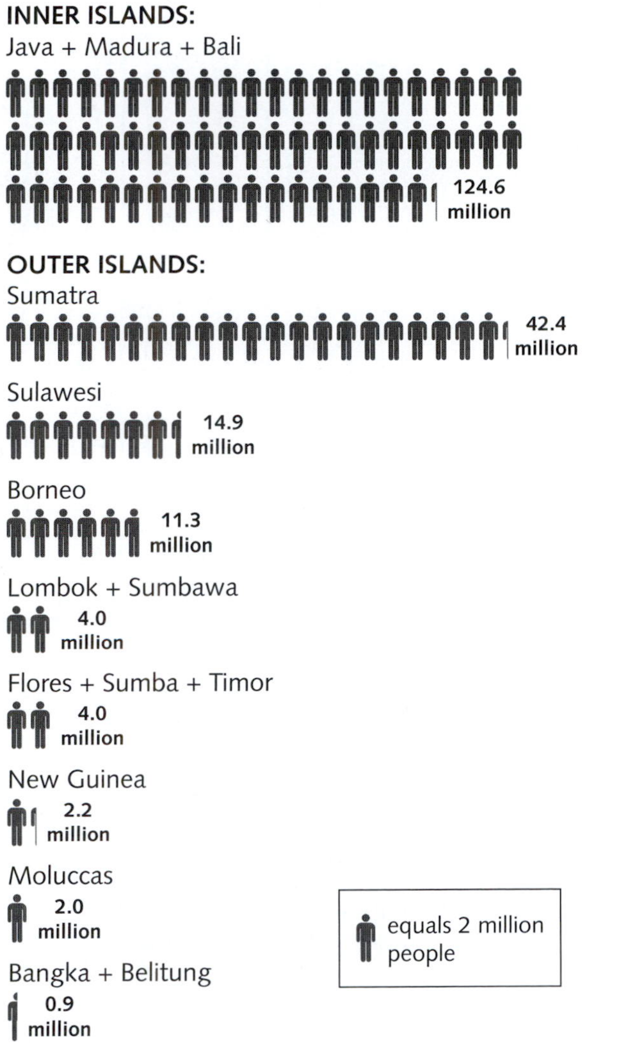

INNER ISLANDS:

Java + Madura + Bali — **124.6 million**

OUTER ISLANDS:

Sumatra — **42.4 million**

Sulawesi — **14.9 million**

Borneo — **11.3 million**

Lombok + Sumbawa — **4.0 million**

Flores + Sumba + Timor — **4.0 million**

New Guinea — **2.2 million**

Moluccas — **2.0 million**

Bangka + Belitung — **0.9 million**

(figure icon) equals 2 million people

Source: Indonesian Census 2000

Constructed Response Use a complete sentence to write the answer in the space provided.

18 Why do you think some islands are listed in groups of two or three?

___

___

Name Class Date

Southeast Asia Today

CHAPTER TEST A

Part 4: Document-Based Question Use the documents and your knowledge of Southeast Asia to answer the questions below.

Introduction

Asian elephants are somewhat smaller than African elephants. Both species are endangered. In the wild, Asian elephants can live up to 60 years. Because they are so large—weighing up to 5.5 tons—elephants need enormous amounts of food to survive. The loss of habitat where they can find enough food is just one of many reasons the species is now endangered.

Objective: Analyze why Asian elephants are endangered.

DOCUMENT 1 Passage from "Ivory Boom in Vietnam Threatens Asia's Last Wild Elephants"

Indochina's few surviving wild elephants are under increasing threat from booming illegal ivory prices in Vietnam, according to . . . TRAFFIC, the wildlife trade monitoring network:

> "An assessment of the illegal ivory trade in Vietnam said Vietnamese illegal ivory prices could be the highest in the world, with reports of tusks selling for up to U.S. $1500/kg [$3,300 per pound] and small, cut pieces selling for up to $1863/kg [$4,098 per pound]," TRAFFIC said in a news statement.

Source: "Ivory Boom in Vietnam Threatens Asia's Last Wild Elephants," *National Geographic News Watch*, January 16, 2009.

Constructed Response Use complete sentences to write the answers in the space provided.

19 Why might the price for ivory be so high in Vietnam?

____

____

20 Why is such trade a threat to elephants?

____

____

DOCUMENT 2 Passage from *National Geographic Today*, October 16, 2002

Thais often say elephants helped build their nation. For centuries they were Thailand's tanks, taxis, and bulldozers. As such, a contradiction developed: These beasts of burden became cultural icons. . . But the elephants' status as cultural icons hasn't stopped a slide to near-extinction in Thailand. The World Conservation Union, based in Gland, Switzerland, lists the Asian elephant as endangered. A century ago, there were 100,000 elephants in Thailand. That number has fallen 95 percent, primarily due to loss of habitat.

Source: Jennifer Hile, *National Geographic Today*, October 16, 2002.

Constructed Response Use complete sentences to write the answers in the space provided.

21 What role have elephants traditionally played in Thai culture?

22 What is the state of the elephant population in Thailand today?

DOCUMENT 3 Chart of Wild Elephant Population in Southeast Asia

ASIAN ELEPHANTS IN THE WILD IN SOUTHEAST ASIA

Country	Estimated Numbers
Cambodia	400–600
Indonesia	1,180–1,557
Laos	781–1,202
Malaysia	2,351–3,066
Myanmar	4,000–5,300
Thailand	3000–3,700
Vietnam	76–94

Sources: Asian Nature Conservation Foundation

Constructed Response Use complete sentences to write the answers in the space provided.

23 In which country is the wild elephant population most endangered?

24 Judging from the chart, is the Asian elephant in serious danger of disappearing from the wild? Explain.

Extended Response Write a paragraph to answer the question. Use information from all three documents and your knowledge of Southeast Asia in writing your paragraph. Use the back of this page or a separate piece of paper to write your answer.

25 Suppose that you are in charge of a publicity campaign to try to save wild Asian elephants. What reasons would you give for saving the elephant and what recommendations would you make?

Name Class Date

Southeast Asia Today

CHAPTER TEST B

Part 1: Multiple Choice Circle the best answer for each question from the choices available.

1 Which Southeast Asian country is mostly Catholic?
- **A** Cambodia
- **B** Myanmar
- **C** Philippines
- **D** Vietnam

2 Which buildings in Thailand have been influenced by designs from India, the Khmer empire, and China?
- **A** royal palaces
- **B** Buddhist temples
- **C** Christian schools
- **D** post offices

3 How did Chinese come to be the dominant language of Singapore?
- **A** colonialism
- **B** migration
- **C** religion
- **D** television

4 What challenge is posed by language diversity in Southeast Asia?
- **A** accessing news on the Internet
- **B** learning culture
- **C** trading with foreign countries
- **D** unifying countries

5 What law was passed in 1975 to try to protect elephants?
- **A** creation of national park
- **B** ban on ivory trade
- **C** tax on deforestation
- **D** outlawing of large guns

6 On what island do more than half of all Indonesians live?
- **A** Bali
- **B** Borneo
- **C** Java
- **D** Sumatra

7 When the Indonesian government promoted relocation, what were officials hoping to spread throughout the islands?
- **A** democratic government
- **B** industrial development
- **C** modern technology
- **D** official language

8 Singapore was once part of which Southeast Asian country?
- **A** Cambodia
- **B** Malaysia
- **C** Myanmar
- **D** Vietnam

9 How did the government of Singapore attract businesses?
- **A** consumer discounts
- **B** minimum wages
- **C** tax incentives
- **D** workplace regulations

10 How does the Malaysian government try to control Malaysian society?
- **A** banning cell phones
- **B** discouraging education
- **C** limiting information
- **D** reducing economic growth

Southeast Asia Today

CHAPTER TEST B

Part 2: Interpret Maps Use the map and your knowledge of Southeast Asia to answer the questions below.

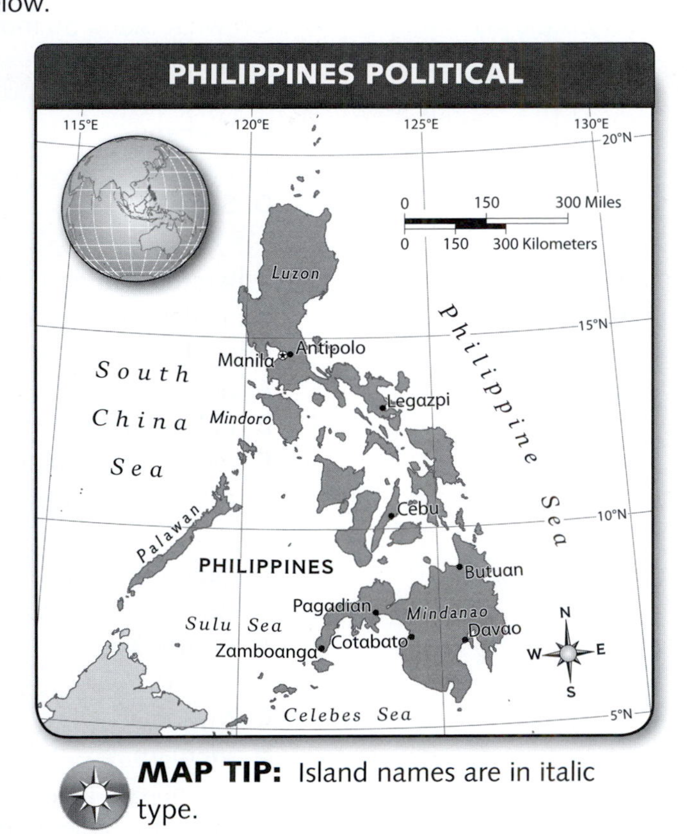

MAP TIP: Island names are in italic type.

11 What is the capital of the Philippines?
- **A** Antipolo
- **B** Cebu
- **C** Davao
- **D** Manila

12 In what section of the country is the capital located?
- **A** north
- **B** east
- **C** south
- **D** west

13 What are the two largest islands?
- **A** Luzon and Mindoro
- **B** Luzon and Mindanao
- **C** Mindoro and Palawan
- **D** Mindanao and Palawan

14 Which island has the highest number of major cities?
- **A** Luzon
- **B** Mindanao
- **C** Mindoro
- **D** Palawan

BLACK-AND-WHITE COPY READY

Name Class Date

Southeast Asia Today

CHAPTER TEST B

Part 3: Interpret Graphs Use the graph and your knowledge of Southeast Asia to answer the questions below.

15 Which two islands have a population between 10 million and 20 million?
- **A** Borneo and Sumatra
- **B** Borneo and Sulawesi
- **C** Java and Sumatra
- **D** Flores and Sulawesi

16 What is the population of Sumatra?
- **A** 2.0 million
- **B** 14.9 million
- **C** 42.4 million
- **D** 12.6 million

17 What island is grouped with Bangka?
- **A** Bali
- **B** Belitung
- **C** Java
- **D** Sumba

18 How does the population of Sumatra compare to the population of the other outer islands?
- **A** smallest of all
- **B** about equal to all the others combined
- **C** third largest
- **D** twice as much as the others combined

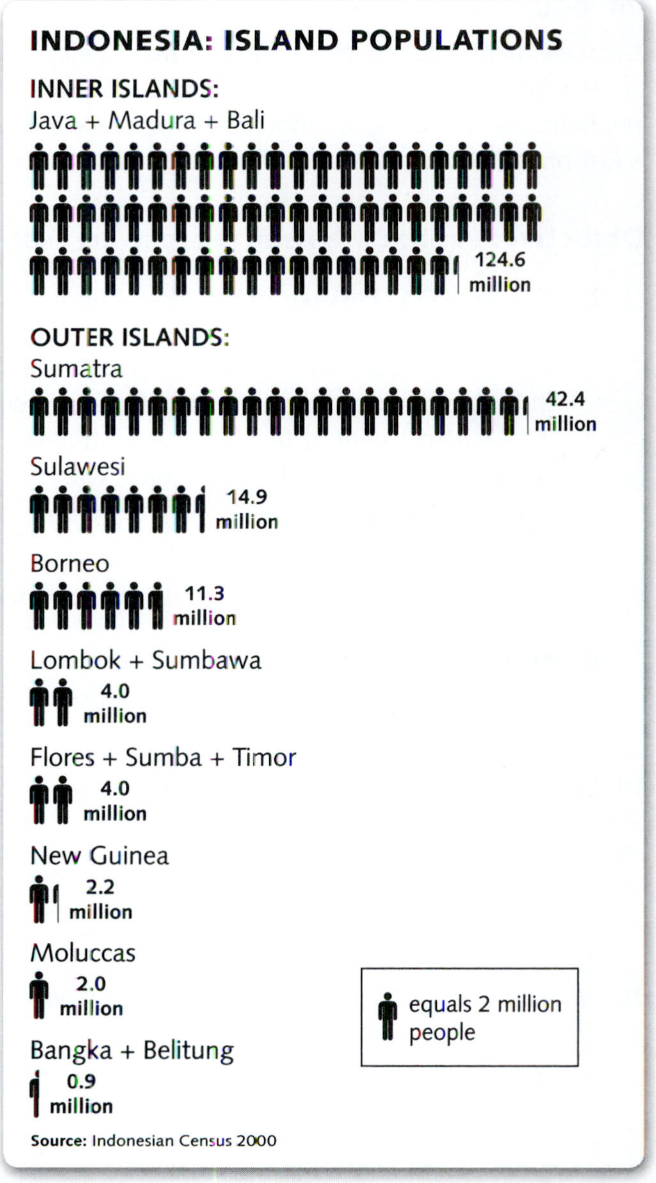

INDONESIA: ISLAND POPULATIONS

Source: Indonesian Census 2000

CHART TIP: Note that some islands are listed individually and some are listed in groups.

Name Class Date

Southeast Asia Today

CHAPTER TEST B

BLACK-AND-WHITE COPY READY

Part 4: Document-Based Question Use the documents and your knowledge of Southeast Asia to answer the questions below.

Introduction

Asian elephants are somewhat smaller than their cousins, African elephants. Both species are endangered. In the wild, Asian elephants can live up to 60 years. Because they are so large—weighing up to 5.5 tons—elephants need enormous amounts of food to survive. The loss of habitat where they can find enough food is just one of many reasons the species is now endangered.

Objective: Analyze why Asian elephants are endangered.

DOCUMENT 1 Passage from "Ivory Boom in Vietnam Threatens Asia's Last Wild Elephants"

Indochina's few surviving wild elephants are under increasing threat from booming illegal ivory prices in Vietnam, according to . . . TRAFFIC, the wildlife trade monitoring network:

> "An assessment of the illegal ivory trade in Vietnam said Vietnamese illegal ivory prices could be the highest in the world, with reports of tusks selling for up to U.S. $1500/kg [$3,300 per pound] and small, cut pieces selling for up to $1863/kg [$4,098 per pound]," TRAFFIC said in a news statement.

Source: "Ivory Boom in Vietnam Threatens Asia's Last Wild Elephants," *National Geographic News Watch*, January 16, 2009.

Constructed Response Write the answer to each question in the space provided. You do not need to write complete sentences.

19 What substance is being traded illegally?

____

____

20 How does this trade affect elephant populations? Explain.

____

____

DOCUMENT 2 Passage from *National Geographic Today*, October 16, 2002

Thais often say elephants helped build their nation. For centuries they were Thailand's tanks, taxis, and bulldozers. As such, a contradiction developed: These beasts of burden became cultural icons. . . But the elephants' status as cultural icons hasn't stopped a slide to near-extinction in Thailand. The World Conservation Union, based in Gland, Switzerland, lists the Asian elephant as endangered. A century ago, there were 100,000 elephants in Thailand. That number has fallen 95 percent, primarily due to loss of habitat.

Source: Jennifer Hile, *National Geographic Today*, October 16, 2002.

Constructed Response Write the answer to each question in the space provided. You do not need to write complete sentences.

21 How do the Thai people view elephants?

____

22 How well protected are elephants in Thailand? Explain.

____

DOCUMENT 3 Chart of Wild Elephant Population in Southeast Asia

ASIAN ELEPHANTS IN THE WILD IN SOUTHEAST ASIA

Country	Estimated Numbers
Cambodia	400–600
Indonesia	1,180–1,557
Laos	781–1,202
Malaysia	2,351–3,066
Myanmar	4,000–5,300
Thailand	3000–3,700
Vietnam	76–94

Sources: Asian Nature Conservation Foundation

Constructed Response Write the answer to each question in the space provided. You do not need to write complete sentences.

23 Which country has the largest population of wild elephants?

____

24 Which two countries have the smallest populations of wild elephants?

____

Extended Response Write a paragraph to answer the question. Use information from all three documents and your knowledge of Southeast Asia in writing your paragraph. Use the back of this page or a separate piece of paper to write your answer.

25 What human activities endanger Asian elephants in the wild, and what has been the result?

GEOGRAPHY & HISTORY

SECTION 1.1 PHYSICAL GEOGRAPHY

Reading and Note-Taking

Title: Physical Geography

Main Idea: Southeast Asia is a mountainous region, with both mainland and island nations.
"Mainland Countries" Details: Mainland countries include Myanmar, Thailand, Laos, Vietnam, and Cambodia. Many people in the region live in small villages in the mountains or near waterways. The river deltas also have dense populations.
"Island Countries" Details: Island countries include Indonesia and the Philippines. These countries sit on the Ring of Fire, and experience a high level of seismic activity.

Map Description: The physical map shows that many parts of the region are isolated from one another. Both island and mainland countries have a number of rivers and mountain ranges.
Typhoon Photo Description: The photo shows a satellite view of an enormous storm, with a spiral form similar to a hurricane.
Tsunami Photo Description: The photo shows and enormous ocean wave crashing on the shore.

Compare and Contrast The Philippines. Because the Philippines is made up of thousands of islands, it has much more coastline, and therefore more exposure to tsunamis. The inland parts of Cambodia are probably safe from the effects of a tsunami.

Vocabulary Practice

Sample Paragraphs: Land bridge and landlocked are both used to describe landforms. Land bridges are narrow strips of land that connect larger landmasses. For instance, land bridges once connected Indonesia and the Philippines with the Asian mainland, but these isthmuses disappeared when sea levels rose more than 6,000 years ago. Landlocked areas are those that are surrounded by land on all sides. In Southeast Asia, only Laos is landlocked, as all other countries have coastlines.

Tsunamis and typhoons are both types of natural disasters that can occur in Southeast Asia. However, tsunamis are produced by seismic activity such as undersea earthquakes. Typhoons, on the other hand, are similar to hurricanes and are a product of the region's tropical climate. These fierce storms often bring high winds and heavy rains during the wet monsoon season.

GeoActivity Compare Past and Present Land Areas

1. Sumatra, Java, Bali, and Borneo were linked. Sulawesi and the Philippine Islands were not linked. The oceans separating unlinked islands were so deep that lower sea levels did not create land bridges there.
2. Possible response: Human cultures and plant and animal species could have moved and spread easily throughout the areas that were connected by land bridges, which would have provided new sources of food and the sharing of ideas.
3. Possible response: Island countries will become smaller and smaller in size until they eventually disappear under the sea level.

SECTION 1.2 PARALLEL RIVERS

Reading and Note-Taking

Mekong River
Location: Myanmar, Laos, Thailand, and Vietnam
Major Cities Along the River: Ventiane, Ho Chi Minh City
Economic Uses: valuable for rice growing, hydroelectricity
Other Notable Features: longest river in Southeast Asia; river delta is the size of West Virginia

Chao Phraya River
Location: Thailand
Major Cities Along the River: Bangkok
Economic Uses: irrigating rice fields; transportation
Other Notable Features: empties into Gulf of Thailand

Irrawaddy River
Location: Myanmar
Major Cities Along the River: Mandalay
Economic Uses: rice farming, transportation
Other Notable Features: delta grows by 165 feet each year; river level rises more than 30 feet during rainy season

Vocabulary Practice

Word: ecologist
Definition: a scientist who studies the relationship between organisms and their environments
In Your Own Words: someone who focuses not only specific animals or plants, but also on the ways these things form an interdependent system
Sentence: The ecologist wanted to collect data to find out whether cutting down hardwood trees in an area had affected the species of birds who nested there.

Word: subsistence fishing
Definition: the system of catching just enough fish to live on
In Your Own Words: fishing only enough to provide food for oneself and one's family
Sentence: Subsistence fishing is a way of life for some people who live near rivers.

GeoActivity Research an Environmental Issue

1. Students should use the research links or other print or online sources that are valid and reliable.
2. Dam building, overfishing, and habitat destruction are major problems identified in the research links.
3. Students' solutions will vary but may include the following: government intervention to put limits on fish catches; government intervention to stop the building of dams; programs such as Zeb Hogan's Megafishes Project, which works to protect large freshwater fish.

SECTION 1.3 THE MALAY PENINSULA

Reading and Note-Taking

Main Idea: Mineral Resources
Detail: tradition of tin mining in Malaysia and Thailand
Detail: Bauxite is mined in the southern part of the Main Range.
Detail: depletion of accessible deposits since the 1970s

Main Idea: extensive rain forest
Detail: Rain forest covers about 40 percent of the peninsula.
Detail: hundreds of types of trees and plants
Detail: animals include elephants, rhinos, and tigers

Main Idea: deforestation
Detail: Some trees have high economic value.
Detail: Forest is cleared to plant specific trees.
Detail: Deforestation causes major environmental problems.

Analyze: Rain forest tourism relies on intact rain forests for the tourists to explore. The growers of palm oil and rubber trees, however, cut down large areas of rain forest for their plantations.

Vocabulary Practice

Word: bauxite
What is it? the raw material used to make aluminum
What is it like? ore, or rock that contains valuable minerals
How does it help you? Understanding the types of valuable mineral resources found in the mountains of the Malay Peninsula helps to understand the economic considerations that must be weighed against environmental concerns.

Word: biodiversity
What is it? the variety of species in an ecosystem
What is it like? biology, diversity
How does it help you? Understanding that the rain forests of the Malay Peninsula have great biodiversity helps to see why many people are concerned with protecting the rain forests.

GeoActivity Graph Global Deforestation Rates

1. Students' graphs should resemble the following:

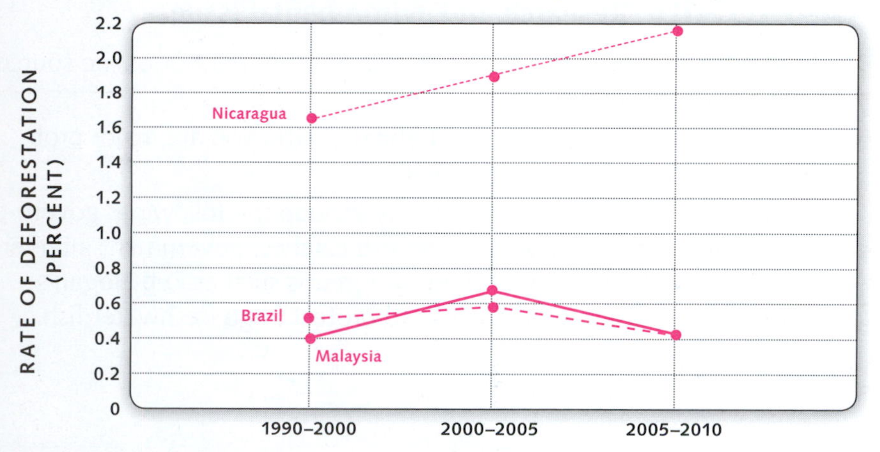

2. 2000–2005; the deforestation rate dropped sharply
3. Deforestation in Nicaragua has increased while it has slowed in the other countries. The percentages in this country are much higher.
4. Students' responses will vary but may include the following ideas: Malaysia and Brazil may be finding new ways to restore forest growth, plant trees, and prevent the cutting of trees for fuel and farming; these countries likely have strong economies, which allows them to focus on environmental issues; Nicaragua might have economic problems, which means people need to clear the land for fuel and the government is unable to focus on environmental issues.
5. Students may suggest stricter laws and enforcement, education and information sharing, or an increased popular concern.

SECTION 1.4 ISLAND NATIONS

Reading and Note-Taking

I. Dynamic Geographic Zone
A. multiple plates collide, creating volcanic mountains
B. seemingly dormant volcanoes may erupt after hundred of years of inactivity

II. Indonesia
A. country is by far the region's largest in terms of both area and population
B. majority of population and most major cities found on island of Java

III. The Philippines
A. similar settlement patterns to Indonesia, with most people on lowland plains
B. many small, rural settlements, sustained by fishing and rice farming

Vocabulary Practice

ACADEMIC VOCABULARY

Sample sentence: There is a lot of agriculture near volcanoes in Southeast Asia because volcanic material enhances the soil.

Comparison Chart

Word: dormant
Definition: inactive for long periods of time

Word: dynamic
Definition: continuously changing or active

Similarities: Both words are used to describe whether or not something is active, and both help to understand the ways seismic activity shapes Southeast Asia's landscape.

GeoActivity Analyze the Effects of Krakatoa

1. Students should either use the provided research links at Connect to NG or find other reliable print or online sources.
2. Possible effects include the following: loss of human life on nearby islands due to tsunamis; loss of animal and plant life nearby and at greater distances from ash, lava, and tsunamis; devastation of the islands of Java and Sumatra; drop in the average global temperature from ash clouds; increased land mass from ash added to nearby islands; half of Krakatoa was blown away.
3. Many habitats were probably destroyed and took a long time to return. People and other animals either starved or had to migrate someplace else.

SECTION 1.5 DISCOVERING NEW SPECIES

Reading and Note-Taking

Central Concept: Zoologists search for new plant and animal species in Indonesia's remote Foja Mountains.

Supporting Detail: Because the Foja Mountains have been isolated from human contact, zoologists believe the area may have many undiscovered species.
Supporting Detail: Expeditions to the region include experts on different plant and animal species.
Supporting Detail: During the 2005 expedition, researchers discovered dozens of new species.
Supporting Detail: Scientists on these expeditions learn the process of identifying and documenting new plant and animal life.

Make Inferences: No. If the mountains were easily accessible, there would probably be people living there, and the people would have made an impression on the ecosystem.

Vocabulary Practice

Sample Blog Entry:

Headline: Undiscovered Wildlife

Date: July 6, 2012

After eight days in the Foja Mountains, I continue to be amazed by the number of species the team discovers each day. Zoologist Kristofer Helgen alone has found nine new species of rodents and other small mammals. My personal favorites so far have been the two new types of wallaby Helgen has discovered. They're so tiny, but the resemblance to their cousins the kangaroos is obvious! When planning for the trip, I found it hard to believe that the area had truly remained pristine through the centuries, but after arriving I quickly understood why. In addition to the incredibly dense rainforest, these mountains create an intimidating physical barrier. Yet because the terrain prevented human settlement, today we have the opportunity to discover and observe flora and fauna that no humans have ever seen before. I will of course continue to update the blog and keep readers informed of any new discoveries.

GeoActivity Investigate New Species

1. Students' charts will vary depending on the additional research they do, but they will most likely include the following information for each species:

   Blossom Bat: nocturnal; the "hummingbird of the bat world"; pollinates plants like a bumblebee

   Bent-Toed Gecko: lives in trees and on the ground; has bent toes that help it grip

   Dwarf Wallaby: size of a rabbit; smallest member of kangaroo family

   Woolly Giant Rat: approximately 3 lbs.; five times the size of a city rat

   Pinocchio Frog: lives mostly in treetops; male has a nose that points up when it makes noise and hangs down when it is silent

   Tree Mouse: moves through the forest on a network of branches and vines; hardly touches the ground

2. Students' opinions will vary but should be supported by reasons and evidence.

SECTION 1 REVIEW AND ASSESSMENT

Vocabulary

1. E	**3.** G	**5.** I	**7.** B	**9.** H
2. A	**4.** J	**6.** F	**8.** D	**10.** C

Main Ideas

11. mainland nations
12. tropical climate
13. tsunami
14. It is used to irrigate rice fields and as a major transportation route.
15. The Mekong River is the longest.
16. The delta is growing by about 165 feet each year.
17. Malaysia
18. In both countries, most people live along lowland plains, and both countries have extremely large capital cities.
19. rain forest
20. The project gives zoologists a rare opportunity to learn how to discover and identify new species.

Focus Skill: Analyze Cause and Effect

21. Glaciers melted and caused the sea level to rise, separating landmasses.
22. The river's depth can vary by more than 30 feet between the rainy season and the dry season.
23. Dams help control water levels, but they also interfere with transportation and can damage the environment.
24. Accessible mineral deposits have been depleted.
25. Possible responses: It can cause landslides; it may drive away tourists hoping to visit the rain forest.
26. The ash creates fertile, nutrient-rich soil, which is ideal for farming.
27. The humid climate is good for growing crops. Some can be grown year-round.
28. The lack of human presence means there are species awaiting discovery that nobody has identified yet.

Synthesize: Answer the Essential Question

A number of geographic factors separate Southeast Asia's countries from one another. Half of the region's countries are found on the Asian mainland, while the other half are island countries. Some of these island countries are made up of hundreds of different islands, with unique cultures found in different areas. The mountainous terrain on the mainland causes many people to live in small villages, although river basins and deltas can be highly populated. Fertile soil in certain parts of the region enables a great deal of farming and leads to high population density, while other areas, such as Indonesia's Foja Mountains, remain isolated.

SECTION 1 STANDARDIZED TEST PRACTICE

Multiple Choice

1. C **2.** D **3.** A **4.** C **5.** B **6.** A **7.** D **8.** B **9.** A **10.** C

Constructed Response

11. The eruption happened soon after the earthquake that set off the tsunami.

12. The volcano had already been building up steam for several days before the earthquake.

Extended Response

13. The volcano and the earthquake's epicenter might lie on the same fault line or along the intersection of the same tectonic plates.

Data-Based Questions

14. The vertical movement would be more likely to force the water column up, producing a tsunami wave.

15. Southeast Asia probably has more fault lines than most other parts of the world. Although tectonic plates collide around the world, Southeast Asia is at the intersection of not just two but four tectonic plates.

SECTION 2.1 ANCIENT VALLEY KINGDOMS

Reading and Note-Taking

Khmer Empire
- centered along Mekong River valley in Cambodia
- built Angkor Wat temple complex in the 1100s

Dai Viet Kingdom
- formed in A.D. 939, after breaking free from China
- women enjoyed higher social standing than in China

Srivijaya
- arose in Sumatra in the A.D. 600s
- controlled trade from South Asia to China

Sailendra
- flourished in Java from about A.D. 780 TO 850
- built Borobudur temple complex

Majapahit
- arose in eastern Java around A.D. 1300
- gained power through controlling trade

Compare and Contrast: Both empires lasted for several centuries. However, the Khmer Empire was on the mainland, while the Srivijaya Empire was on an island. Also, the Khmer Empire built a great Hindu temple, while the Srivijaya Empire was known as a center of Buddhist study.

Vocabulary Practice

Students' drawings will vary, but should look similar to the photograph of Angkor Wat in Section 2.1.

Sample paragraph:

Here at the historic temple of Angkor Wat in Cambodia, we can marvel at ancient monumental architecture and examine countless elaborate bas-relief sculptures that date back 900 years. Visitors are often overwhelmed by the elaborate Hindu-inspired sculptures that cover the walls of the enormous temple complex. Learn about the history of the ancient Khmer Empire, and investigate efforts to preserve these historic structures. Visitors to Southeast Asia will not want to miss a chance to connect with history at Angkor Wat.

GeoActivity Solve a Puzzle About Ancient Kingdoms

Across

1 SAILENDRA
2 CHINA
4 DAI VIET (written as one word)
7 ANGKOR WAT (written as one word)
8 BUDDHISM
9 KHMER

Down

1 SRIVIJAYA
3 MAJAPAHIT
5 BOROBUDUR
6 BAS-RELIEF (written as one word with no hyphen)

Mystery Word
CAMBODIA

SECTION 2.2 TRADE AND COLONIALISM

Reading and Note-Taking

Topic: Colonialism in Southeast Asia

Introduction: A number of European powers struggled for control in Southeast Asia, attracted by the region's valuable spices.

Spanish colonies
Details: Spain claimed the Philippines in the 1500s and retained control of the islands until 1898.

Dutch colonies
Details: The Dutch East India Company controlled much of the region, particularly in present-day Indonesia.

British colonies
Details: Arriving in the region long after Spain and the Netherlands, Britain took control of much of the Malay Peninsula and present-day Myanmar.

French colonies
Details: The French controlled much of Indochina in the 19th century but agreed to allow Siam to remain independent.

Conclusion: Recognizing the value of the region's natural resources, European powers claimed much of Southeast Asia and caused major changes to the countries in the region.

Vocabulary Practice

Word: colonialism

1. Colonialism—one country ruling and developing trade in another country for its own benefit—continued in Southeast Asia well into the 20th century.
2. the system in which one country takes control of other countries with the expectation or goal of benefiting economically
3. For centuries, colonialism led Europeans to exert power and exploit local populations in Asia, Africa, the Americas, and the Pacific Realm.
4. The suffix "-ism' indicates that this is a system or theory. Other examples of words with this suffix include fascism, communism, imperialism, and pacifism.

Word: monopoly

1. European influence arrived in the 1500s, as merchants hoped to establish a monopoly, or complete control of the market, of the spice trade.
2. the condition when one company has total control over the supply of a good, with no competition
3. The cable company had a monopoly over the local market—no other company was allowed to offer cable services in the area.
4. The prices of goods under a monopoly increases, as there is no competition offering to sell the same goods for less money.

GeoActivity Map the Spice Trade

1. Students' maps should resemble the following:

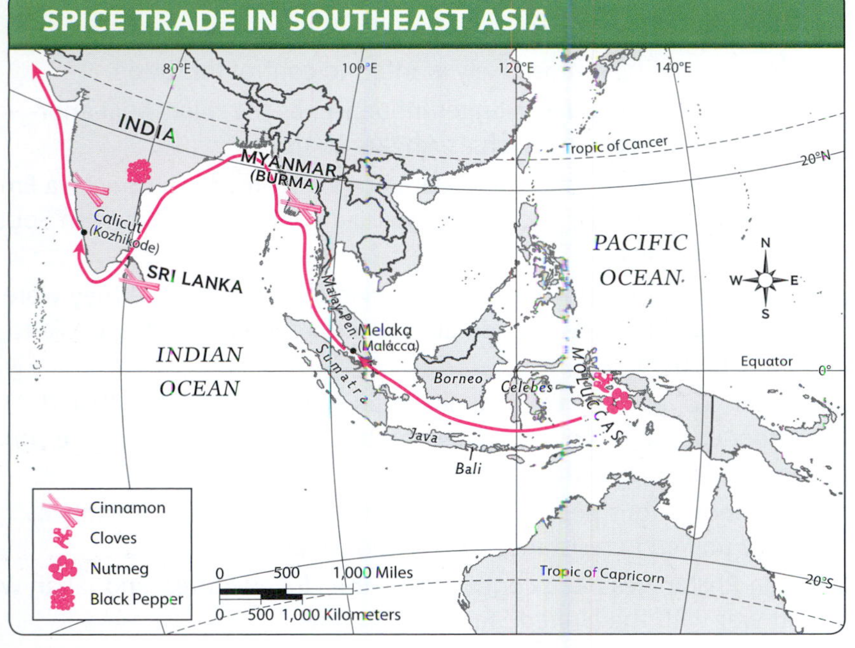

2. These islands were the only source of several valuable spices.
3. Students' responses will vary, but students may speculate that European traders began to plant these spices in other tropical areas they controlled, such as Africa and the Caribbean islands.

SECTION 2.3 INDONESIA AND THE PHILIPPINES

Reading and Note-Taking

Indonesia:

- possible home to earliest humans, as suggested by fossils
- early Indonesians traveled by sea as early as 2500 B.C
- known to Europeans as the Spice Islands
- controlled by the Dutch until World War II

The Philippines:

- controlled by the Spanish beginning in the 1500s
- Trade with China led to many Chinese settlers.
- As Spain lost power, Manila was opened to trade with other nations.

Both:

- focus for European colonialism
- Independence movements began in late 1800s or early 1900s.
- controlled by Japan during World War II
- Today, both are independent nations.

Spain's control over the Philippines diminished in the 1800s, while the Dutch grew more powerful and wealthy from their actions in Indonesia.

Vocabulary Practice

1. The Chinese settlers in the Philippines were important businesspeople, conducting a great deal of commerce in the nation.
2. In some areas, commerce is dominated by large corporations.
3. commercial, commercialized
4. Possible responses: dinosaurs, mammoths
5. Artifacts are objects made by humans, whereas fossils are the remains or impressions of living beings.
6. Fossils can show where people were living and when.

GeoActivity Analyze Achievements of Emilio Aguinaldo

1. **Obstacles:** sent into exile in Hong Kong after the Katipunan revolt; Philippine independence not recognized by Spain or the United States; captured by U.S. forces during the Philippine-American War; lost presidential election in 1935

   Achievements: upon return from exile, declared independence from Spain and was elected president; organized a revolution against the United States; saw the Philippines gain independence in 1946; worked to promote nationalism and democracy
2. Possible response: The United States wanted the Philippines as a colony because of its unique resources and its location as an important trading center in the region. Also, the United States was interested in becoming a world power and wanted to rival European imperial powers with its own colonial possessions.

SECTION 2.4 THE VIETNAM WAR

Reading and Note-Taking

Possible responses:

Document 1
Who wrote or created this document? President Lyndon Johnson
What is the main idea of this document? The United States is determined not to leave Vietnam. If the U.S. did retreat, the forces of communism would have to be fought in other countries.
What is the point of view of the writer or person shown? The United

States has a responsibility to the rest of the world to fight communism in Vietnam.
Why do you think the person feels this way? He believes that the spread of communism would be bad for other countries.

Document 2
Who wrote or created this document? Ho Chi Minh
What is the main idea of this document? The North Vietnamese will not negotiate a peace with the United States and will continue to fight for unification and independence for Vietnam.
What is the point of view of the writer or person shown? He believes the United States is the aggressor and wants to rule Vietnam as a colonial power.
Why do you think the person feels this way? Southeast Asia has a history of being dominated by colonial powers from the West.

Document 3
Possible responses: The photo shows a soldier walking through mud in a rain forest setting. The soldier is wet to the thighs and having a hard time slogging through the mud, giving a good idea of how hard it was to walk or fight over the ground in Vietnam.

Vocabulary Practice

ACADEMIC VOCABULARY

Sample sentence: The United States government feared that if South Vietnam fell, other countries would as well, and the entire region could be transformed into a string of Communist states.

Word: launch
Definition: to start something or set it in motion
Characteristics: giving something a dramatic or impressive start, such as to launch a campaign
Example: launching a sailboat, launching a rocket, launching an ad campaign
Non-Example: stall, stop

Word: resistance
Definition: opposition, or a group that rebels or opposes
Characteristics: Resistance can mean one's ability to withstand something, such as resistance to cold.
Example: the resistance of an underground military organization against an occupying army
Non-Example: acceptance, assistance

GeoActivity Compare and Contrast Two Wars in Asia

1. Possible responses include the following:
   Korean War: international forces came to aid South Korea; UN forces recapture Seoul from North Korea; North and South Korea remain divided; American troops still guard the DMZ
   Vietnam War: primarily U.S. forces came to aid South Vietnam; Saigon falls to North Vietnam; Vietnam reunited as a communist country; no U.S. involvement in the present day
   Similarities: peace talks but no official peace treaty to end war; both wars between Communists and non-Communists; no U.S. victory
2. The major difference is that the Korean War resulted in a split country, whereas North and South Vietnam were united.

SECTION 2 REVIEW AND ASSESSMENT

Vocabulary

1. B **3.** D **5.** C **7.** G
2. F **4.** H **6.** A **8.** E

Main Ideas

9. These waterways connected the Pacific Ocean and the Indian Ocean.
10. Women in Vietnam had higher social standing.
11. The Srivijaya Empire formed before either the Sailendra or the Majapahit.
12. Each level represents a step toward Buddhist enlightenment.
13. the Dutch East India Company
14. These were the only places not under foreign control at the time.
15. Fossils found on Java suggest people may have lived there 1.7 million years ago.
16. Dutch control expanded, and Dutch oppression of the indigenous population increased.
17. Some Filipinos gained wealth and influence through trade with other nations.
18. The U.S. government feared that if South Vietnam fell, communism would spread to other nations.

Focus Skill: Make Inferences

19. Vietnam provided a valuable link to trade with India and other areas, and China most likely wanted to control this trade.
20. India probably had a stronger influence, as the rulers built enormous Hindu temples such as Angkor Wat.
21. Control over the Strait of Malacca probably made the Srivijaya Empire quite wealthy, because they could control trade between South Asia and China.
22. These spices must not have been plentiful in Europe, as they were very valuable there and led European traders to Southeast Asia to acquire them.
23. Japan lost all of its colonial possessions after World War II, and other countries may not have had the resources to try to take control of Indonesia after the war.
24. The indigenous people developed a sense of national identity as they unified in resistance to the colonial powers.
25. The Philippines were controlled by the United States, and Japan was at war with the United States at this time.
26. Success for South Vietnam must have seemed highly unlikely at the time the U.S. forces withdrew in 1973.

Synthesize: Answer the Essential Question
Southeast Asia's multiple physical barriers have had a strong influence on the region's history. The division of the region into many islands and peninsulas enabled the simultaneous existence of many kingdoms, each of which could use the region's waterways to grow rich through trade. However, these same factors made the region vulnerable to European colonialism, and for several centuries, European powers controlled most of the region. Because of Southeast Asia's strategically valuable location and broader influence, several countries in the region became the source of conflict during major wars in the 20th century.

SECTION 2 STANDARDIZED TEST PRACTICE

Multiple Choice

1. B 3. D 5. D 7. B 9. A
2. A 4. C 6. A 8. D 10. B

Constructed Response

11. Concealed means hidden from view, as the temple was difficult or impossible to see through the thick forest.
12. No, the temple was probably meant to be imposing and visible from a distance and became concealed over the course of several centuries.

Extended Response

13. The temple is probably in Cambodia. We know that it is a Khmer temple, and the Khmer Empire was centered along the Mekong River in Cambodia. We can assume this is a Hindu temple, as the Khmer Empire built other Hindu temples, such as Angkor Wat.

Data-based Questions

14. The architecture is very elaborate, decorate, and complex.
15. After armies from modern-day Thailand conquered the city, Angkor Wat fell into ruin. Given the region's tropical climate, it probably did not take very long for the elements to destroy these original wooden structures.

FORMAL ASSESSMENT

SECTION 1 QUIZ

1. B 3. A 5. D
2. C 4. D 6. A

7. Melting glaciers caused the sea level to rise, covering up land bridges and creating separate islands.
8. Four tectonic plates come together in Southeast Asia, and their movement often causes volcanoes.

SECTION 2 QUIZ

1. B 3. C 5. C
2. C 4. B 6. D

7. Hinduism and Buddhism
8. Because of the region's location and resources, over the centuries many countries and cultures have traded in and colonized the region, impacting its religion, culture, government, and economy.

GEOGRAPHY & HISTORY TEST A

Part 1: Multiple Choice

1. B 3. A 5. B 7. A 9. C
2. A 4. C 6. B 8. C 10. A

Part 2: Interpret Maps

11. A 12. B 13. C

14. Indonesia's most populated island is located around 10° south of the equator.

Part 3: Interpret Charts

15. C 16. D 17. B

18. The islands are home to ecoregions that support many unique plant and animal species.

Part 4: Document-Based Question
Constructed Response

19. They were cleared in order to plant more of the trees that had significant economic value.
20. It resulted in deforestation, which causes landslides.
21. Possible answer: The government wants the country to grow economically and modernize; the government wants the country to have a role in the global economy.
22. Possible answer: One might infer that the government will encourage the efforts of corporations in general and that the needs of indigenous communities for land and their traditional way of life might be compromised in order for economic growth and prosperity to occur.
23. The lake is a sanctuary for many freshwater species; researchers do extensive work finding out more about these species; and it has become the first UNESCO designated biosphere reserve in the country of Malaysia.
24. The information about handicraft production in the lake area points to a potential conflict between efforts to keep freshwater species alive and healthy and efforts to develop industries that have economic potential but that also might pollute nearby waters.

Extended Response

25. Possible answers: Yes, it is possible to strike a balance if the government, companies, and local citizens work together to reach compromises. Or no, unfortunately, balances like this are hard to reach because they require each side to give up something that they truly care about.

GEOGRAPHY & HISTORY TEST B

Part 1: Multiple Choice

1. D 3. D 5. C 7. B 9. D
2. D 4. A 6. A 8. B 10. A

Part 2: Interpret Maps

11. C 12. D 13. B 14. D

Part 3: Interpret Charts

15. B 16. A 17. C 18. A

Part 4: Document-Based Question
Constructed Response

19. Teak wood had great economic value, forming a large part of Thailand's economy.
20. Clearing the forest resulted in landslides.
21. Transportation symbols: light-rail, airline jet, commercial ship; communication symbols: high-rise tower and satellite
22. Possible answer: The government wants the country to grow economically through advances in modern transportation and communication.
23. They do work to find out information about the many freshwater species who live there.
24. Some hope that handicraft production such as textiles, which currently takes place in the area, will grow into a major economic activity for the area.

Extended Response

25. Possible answers: Yes, Malaysia is a country rich with native species and indigenous populations, and to lose either for the sake of developing the economy would be a serious loss. Or no, Malaysia needs to concern itself with the economic welfare of its people rather than thinking the natural environment and native lands are priorities.

TODAY

SECTION 1.1 RELIGIOUS TRADITIONS

Reading and Note-Taking

Animism
- the traditional religion in Southeast Asia
- belief that spirits exist in plants, animals, and objects
- practiced by many small, tribal groups
- rituals performed to please spirits and bring good fortune

Buddhism
- brought to Vietnam by China after 111 B.C.
- Indian traders and pilgrims brought Buddhism to other parts of the region.
- most prominent in mainland countries today

Hinduism
- brought by people from India in the A.D 100s.
- influence found in Cambodian temple at Angkor Wat
- main religion on Bali, an island in Indonesia

Islam
- introduced to the region by Arab traders in the 1300s
- spread from Indonesia to Malaysia
- Indonesia is now the world's most populous Muslim nation.

Christianity
- Spain brought Roman Catholicism to the Philippines in the 1500s.
- France introduced Catholicism to the mainland in the 1700s.
- chief religion in the Philippines and East Timor

Vocabulary Practice

ACADEMIC VOCABULARY

Sample sentence: One of the predominant cultural traditions in our region is outdoor theater and music.

Word Maps

Word: prehistoric
What the Word Means: living or happening before things were written down
What It Is Like: ancient, pre-history, historical
Sentence: It's impossible to know exactly how a religion with prehistoric roots started.

Word: ritual
What the Word Means: something that people do over and over, the same way
What It Is Like: ceremony, religious observance
Sentence: Rituals show how people honor their faith and engage in religious practice.

GeoActivity Map Religion in Southeast Asia

1. Students' maps should resemble the following:

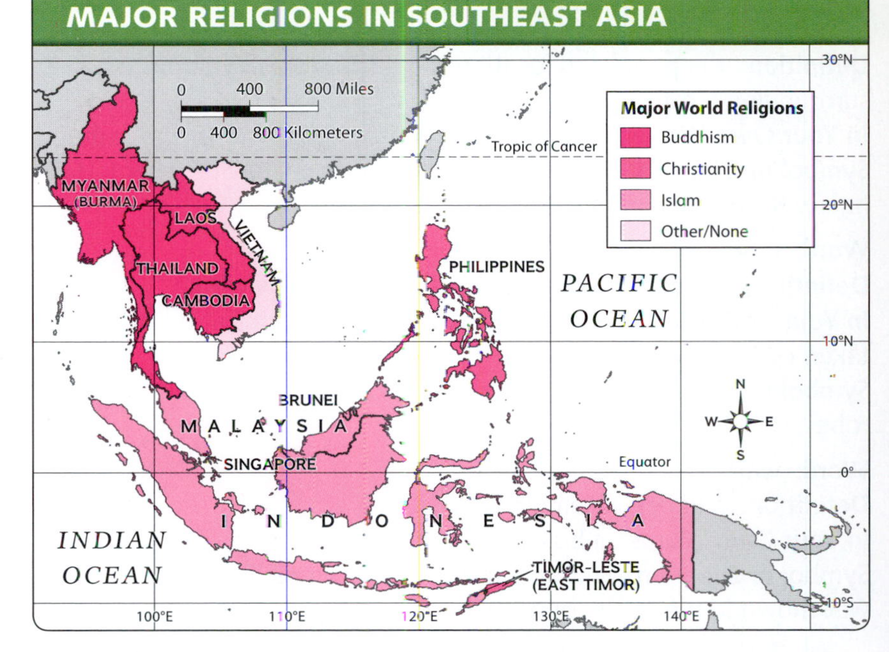

2. Cambodia, Myanmar, and Thailand—Buddhism; Indonesia—Islam; Philippines and Timor-Leste—Christianity
3. Vietnam is a communist country and communist views are traditionally anti-religious. People may still hold private religious beliefs, but they might not publicly say so.

SECTION 1.2 THAILAND TODAY

Reading and Note-Taking

Main Idea: Thai culture today reflects traditional foundations and modern influences.

Detail: Traditional Thai buildings had steeply slanted roofs and were built on legs to withstand floods.
Detail: Buddhist temples known as wats demonstrate Indian, Khmer, and Chinese influences.
Detail: Buddhist monks live simply and focus on meditation and other rituals.
Detail: Younger men are making shorter religious commitments as they migrate away from rural communities.
Detail: Most young people now work in cities, especially Bangkok.
Detail: Urban life and modern conveniences contrast with cultural traditions.

More young people are moving to cities such as Bangkok for school and work. The clothing in cities is Western, and many meals are bought premade rather than home-cooked. Entertainment includes the use of the Internet.

Vocabulary Practice

Word: attribute
Definition: a specific quality or feature
In Your Own Words: the characteristics that help understand or identify something
Symbol or Diagram: Students may draw a steeply slanted roof, an

attribute of traditional Thai architecture.

Word: metropolitan area
Definition: the populated location that includes a city and the surrounding territory
In Your Own Words: a city and the nearby suburbs or neighborhoods
Symbol or Diagram: Students may draw a map showing a central city hub with surrounding neighborhoods.

Word: monk
Definition: a man who devotes himself to religious work
In Your Own Words: a member of a religious organization who has taken certain vows
Symbol or Diagram: Students may draw a traditional Buddhist monk in robes.

Word: wat
Definition: a Buddhist temple
In Your Own Words: a building for Buddhist worship
Symbol or Diagram: Students may create a simple line drawing of the wat shown in the lesson or any similar building.

GeoActivity Graph Thailand's Population Trends

1. All three populations (rural, urban, and total) increased between 1969 and 2009.
2. **Rural Population:** 1969—79 percent; 1979—74 percent; 1989—71 percent; 1999—69 percent; 2009—66 percent
   Urban Population: 1969—21 percent; 1979—26 percent; 1989—29 percent; 1999—31 percent; 2009—34 percent
3. Students' graphs should resemble the following:

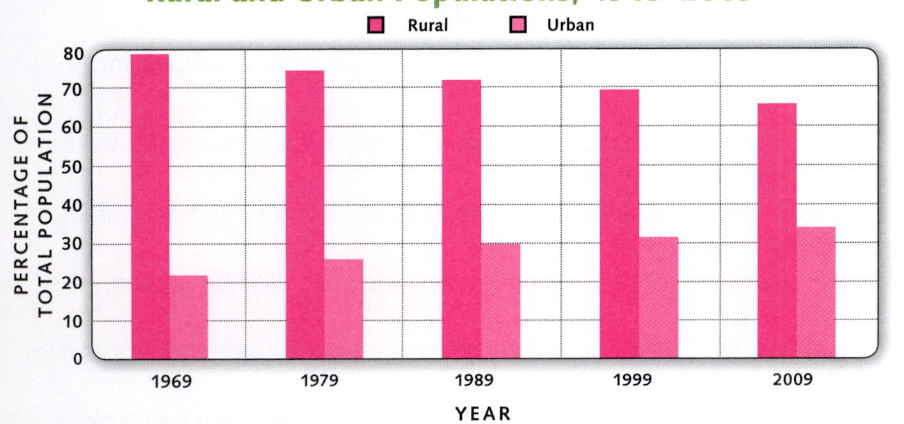

4. The percentage of the population in rural areas decreased while the percentage of the population in urban areas increased.
5. The economy shifted from a reliance on agriculture to a reliance on industry. There were probably fewer jobs in agriculture, which is located in rural areas. There were more jobs available in industry, which is located in urban areas. More people probably moved from rural to urban areas in order to get jobs in the growing industrial sector.

SECTION 1.3 REGIONAL LANGUAGES

Reading and Note-Taking

Official Languages
- An official language is the dominant native language.
- It's used in government, business, education, and the media.
- It usually reflects the name of the country or the largest ethnic group.
- Example: The official language of Laos is Lao.
- It may unify fragmented regions.
- Example: Bahasa has helped unify Indonesia and Malaysia. Or: Diversity of minority languages makes unifying Myanmar more difficult.

Dialects
- These are regional variations of main languages.
- They are often spoken by small groups in isolated communities. Some are not written down.
- Dialects are living languages. But many dialects and ethnic languages are at risk of disappearing.

Non-native Languages
- The need for a common language for traders led to the spread of languages across Southeast Asia.
- With immigration, non-native languages come to a region.
- Example: Chinese and Indian languages spoken in many parts of the region.
- Colonization also brought more languages. European languages continue to be spoken as second languages by many.
- Example: Western languages spoken in the region include English, French, and Dutch.

Vocabulary Practice

Word: adapt
Definition: to adjust to new conditions
Characteristics: a deliberate change based on necessity
Example: to learn the local language after moving to a new place
Non-Example: continuing to use a typewriter even though computers are much more efficient to write with

Word: dialect
Definition: a regional variation of a main language
Characteristics: often spoken by small groups, likely develop over many generations
Example: Cajun French
Non-Example: any official language used by a government and taught in schools

Word: language diffusion
Definition: the spread of a language from its original home
Characteristics: likely resulting from trade or conflict between groups
Example: Spanish being spoken throughout much of Latin America
Non-Example: any dialects that is spoken only by small groups and at risk of disappearing

GeoActivity Analyze Language Relationships

1. Bahasa Malaysia and Bahasa Indonesia are more closely related; they are both part of the Malay subgroup. Khmer and Vietnamese are part of the same language family, but they belong to different subroups.
2. Possible response: As people moved to different islands in the region and became isolated from each other, different versions of a common language developed and these different versions continued to evolve into new languages.

SECTION 1.4 SAVING THE ELEPHANT

Reading and Note-Taking

Cause #1: Poachers kill male elephants for their ivory tusks.
Cause #2: Human populations have increased where the elephants live.
Cause #3: Elephants need a large rain forest habitat, which is being lost.
Cause #4: Humans clear land for logging and mining.
Cause #5: Humans also clear land for raising crops
Cause #6: Farmers kills elephants for eating the crops.

Identify Problems and Solutions: Interactions between humans and elephants can be dangerous for both, so it is crucial to preserve enough elephant habitat so that these interactions are rare.

Vocabulary Practice

Word: domesticating
Definition: teaching an animal how to behave and to obey commands
Explanation: Elephants have been trained to work with humans, providing either services or entertainment.

Word: poaching
Definition: hunting an animal despite laws banning it
Explanation: Although laws passed decades ago banned ivory trading, poaching continues to threaten Southeast Asia's elephants.

Word: restoring
Definition: returning something to its former condition
Explanation: High-tech and low-tech conservation efforts have been made in attempts to restore elephant populations.

GeoActivity Investigate Endangered Species

Students' graphic organizers will vary depending on the species students chose to research.

Possible response: A species' habitat might fall within the borders of several countries, so it would be helpful if the governments of those countries worked together.

SECTION 1 REVIEW AND ASSESSMENT

Vocabulary

1. C
2. E
3. H
4. G
5. D
6. B
7. A
8. F

Main Ideas

9. The Chinese brought Buddhism when they conquered Vietnam in 111 B.C.
10. Islam
11. Buddhism
12. Buildings built on legs or stilts keep them high off the ground.
13. cities or metropolitan areas
14. The people have had less contact over the years with speakers of the dominant language.
15. It served as a common language for traders from China and Arab countries.
16. Trade brings new languages into a country, as people learn the languages to do business.
17. People have cleared the land for farming, mining, and logging.
18. Solar-powered electric fences keep elephants confined to protected places.

Focus Skill: Analyze Cause and Effect

19. Animists believe their rituals will please spirits and bring good fortune on their families or villages.
20. If the beliefs of the ruling power changed, the country's predominant religion also changed.
21. The designs of wats were based on influences from these cultures.
22. Possible responses: Most people wear Western-style clothing, eat prepared foods, and enjoy television, the Internet, and other modern conveniences.
23. The diversity of languages has made unifying the country more difficult.
24. English is used as the common language of global business.
25. The larger human population increasingly threatens the elephants.
26. Elephants stay away, thinking that humans are in the fields.

Synthesize: Answer the Essential Question
Southeast Asian cultures today reflect a number of different influences as well as many centuries-old traditions. Contact with other cultures brought several religions to the region over the centuries, yet many small, tribal groups continue to practice animism. Like religions, languages vary from country to country, and often vary within countries. Similarly, regional dialects may differ considerably from official languages. Interactions with people from East Asia, South Asia, and Europe have brought languages from these places to the region. In Thailand and other countries, people preserve cultural traditions, such as young men serving as monks for a short time, even as modern conveniences become increasingly popular.

SECTION 1 STANDARDIZED TEST PRACTICE

Multiple Choice

1. B
2. B
3. C
4. A
5. B
6. C
7. B
8. D
9. A
10. D

Constructed Response

11. A very large proportion of the Muslims across the world live in Indonesia.
12. Indonesia is "tucked away in a far corner of the world map," so many readers may not know very much about it.

Extended Response

13. The region has a wide diversity of religious traditions, and conflicts may develop between people of different faiths.

Data-based Questions

14. Indonesia's Muslim population is about three times as large as Iran's.

15. Indonesia. The very large Muslim population of Indonesia seems to indicate that Indonesia had the most contact with those who brought Islam to the region.

SECTION 2.1 GOVERNING FRAGMENTED COUNTRIES

Reading and Note-Taking

I. Indonesia's struggle for unity

A. 17,000 islands with more than 300 ethnic groups

B. Java is very urbanized, while Sumatra is rural and agricultural.

II. Malaysia's divisions

A. nation includes territory on Malay Peninsula and island of Borneo

B. tensions between Malay majority and ethnic minorities eased by economic growth

III. Diversity of the Philippines

A. wide variety of ethnic groups and nationalities

B. Tagalog language helps unify population

Evaluate: Possible response: The Philippines seems to have avoided many of the problems confronting Indonesia and Malaysia, because Filipino culture consists of a blend of many different ethnic influences.

Vocabulary Practice

Word: fragmented country
Definition: a nation that is physically divided into separate parts
Detail: Indonesia's thousands of islands create challenges for achieving national unity.
Detail: Malaysia is divided between the Malay Peninsula and the island of Borneo.

Word: motto
Definition: a short saying or expression that guides an individual, an organization, or a nation
Detail: Indonesia's motto is "diversity in unity."
Detail: The Indonesian government hopes the motto will help create a sense of nationhood.

GeoActivity Analyze Remittances and GDP

1. GDP and remittances both increased.
2. 1979: 2 percent; 1989: 3 percent; 1999: 9 percent; 2009: 12 percent
3. Students' graphs will vary slightly based on the scale they chose for the y-axis. They should resemble the following:

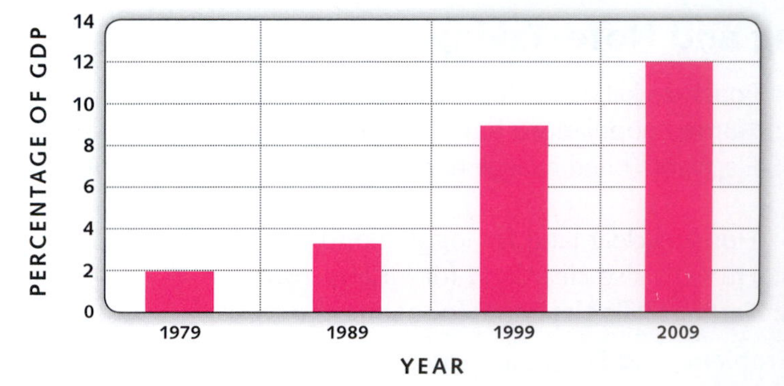

4. The percentage of GDP made up by remittances increased.
5. Possible response: No, it is not a good idea because it signals that there is a problem keeping people employed within the country if more and more money is coming from outside of the country. Remittances are not a good long-term solution to economic stability.

SECTION 2.2 MIGRATION WITHIN INDONESIA

Reading and Note-Taking

Topic: internal migration

What: The Dutch began relocating people from the inner islands to the outer islands.
Why: They hoped this would help unify the vast chain of islands.

What: The government wants to spread the official language throughout the outer islands.
Why: A common language can help unify a fragmented country.

What: Java and Bali remain crowded.
Why: Many settlers have returned to the inner islands, and others have moved there to flee rural poverty.

Analyze Cause and Effect: The difficulty of farming or of finding other work on the outer islands serve as push factors driving migration towards Java.

Vocabulary Practice

1. Sample sentence: The family decided to relocate to a new city when the mother found a new job.
2. Indonesia's government wanted to relocate people from Java's crowded cities to the less densely populated outer islands.
3. Many people from the outer islands have relocated to Java and Bali in search of work.
4. Sample sentence: Fashion trends tend to change from one season to the next.
5. The trend of people moving to the inner islands has worsened the crowding in Indonesia's largest cities.
6. Sample sentence: The government's attempts to relocate people to the outer islands have not been successful, as internal migration trends have largely shown movement from the outer islands to Java and Bali.

GeoActivity Evaluate Internal Migration

Students' charts will vary but should contain some of the following information:

New Settlers: Benefits—more living space on the outer islands; Drawbacks—conflict with native population, damage to the environment, had trouble supporting themselves

Native Population: Benefits—learn the country's official language from new settlers; Drawbacks—conflict with new settlers and challenges to traditional way of life

Populations of Java and Bali: Benefits—less-crowded conditions; Drawbacks—more people ended up settling on the islands to find work

Indonesian Government: Benefits—spread of Indonesian language to help unify the country; Drawbacks—not very effective because people moved back to the highly-populated islands

1. The overcrowded conditions on the inner islands are the strongest push factor. The promise of more space and available farmland are the strongest pull factors.
2. This deforestation would have caused habitats to be destroyed, which would have drastically changed the ecosystem.
3. Standards of living in the cities must be higher than in rural areas if people continue to move from the outer islands to the cities.

SECTION 2.3 SINGAPORE'S GROWTH

Reading and Note-Taking

1. Singapore's location made it an ideal port city.
2. After gaining independence, Singapore thrived as a regional hub for trade.
3. Lee Kuan Yew led the drive to industrialize.
4. Per-capita GDP was more than twice that of Malaysia by the early 2000s.
5. Government leaders invest in infrastructure and high technology.
6. The need for skilled workers led to a strong emphasis on education.

Vocabulary Practice

ACADEMIC VOCABULARY

Sample sentence: Singapore's advanced infrastructure creates the potential for attracting many foreign companies.

industrialize: to develop manufacturing on a wide scale

Sample sentence between *industrialize* and *multinational corporation:* As Singapore industrialized and developed an advanced infrastructure, it attracted many multinational corporations.

multinational corporation: a large business that has operations in many different countries

Sample sentence between *multinational corporation* and *port:* Multinational corporations import and export vast quantities of goods through Singapore's thriving port.

port: a town or city with a harbor where ships can exchange cargo

Sample sentence between *port* and *industrialize:* The wealth accumulated through its success as a port city likely provided Singapore with the resources needed to industrialize.

GeoActivity Graph Singapore's Economic Rise

1. Students' graphs should resemble the following:

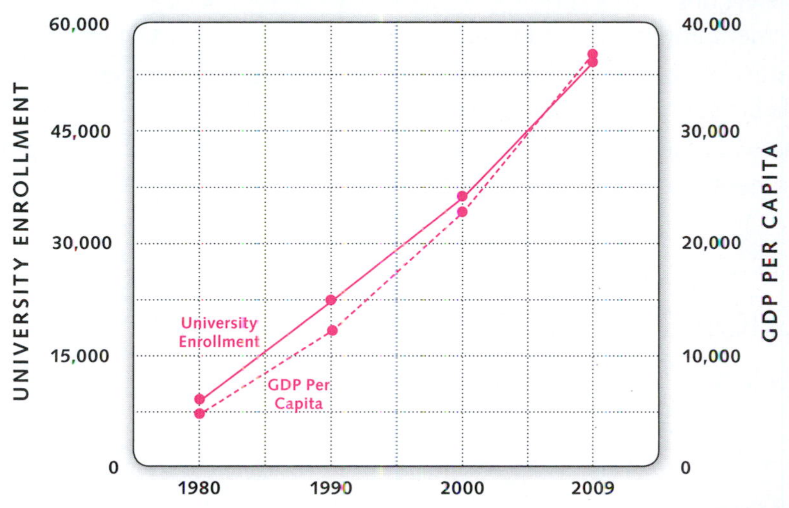

2. By 2009, university enrollment was more than six times what it was in 1980, growing more than three times as fast as the overall population.
3. Possible response: People with a higher level of education probably earn more money, which could explain the relationship between the two factors. Also, people who earn more money are able to afford more education for themselves and their children.

SECTION 2.4 MALAYSIA AND NEW MEDIA

Reading and Note-Taking

Topic: New media in Malaysia

Event 1: For decades, newspapers, televisions, and radios have operated under severe government restrictions.
Event 2: Web sites on the Internet are much freer to criticize the government, although there are some risks involved.
Event 3: Access to information online may have helped sway Malaysian elections in 2008.

Prediction: The Malaysian government will not be able to control information in the way that it has for decades, and this will bring about more changes in the government.

Vocabulary Practice

WDS triangles:

W: emergence
D: the process of coming into being or arriving
S: The emergence of new media has made it increasingly difficult for repressive governments to control the flow of information.

W: reliable

D: trustworthy or dependable
S: People of different generations in Malaysia disagree about which news sources are most reliable.

GeoActivity Explore Effects of New Media

1. Voters have only been able to get information that is in favor of the government, so the government was able to influence the outcome of elections. The Internet gave people a way of criticizing the government and influenced voters to vote against the ruling party.
2. Possible response: Information spread online might not come from a reliable source and could not be entirely factual.
3. The continued development will make it much more difficult for governments to control the spread of information.

SECTION 2 REVIEW AND ASSESSMENT

Vocabulary

1. C
2. F
3. A
4. E
5. B
6. H
7. G
8. D

Main Ideas

9. Sumatra is rural and contains large plantations; Java is urban and has a large population.
10. the Malay Peninsula and part of the island of Borneo
11. The widespread use of Tagalog has helped unify Filipinos.
12. the inner islands
13. The practice of relocating people began before independence.
14. The program has led to conflict rather than national unity, and the inner islands remain crowded.
15. People from the outer islands have moved to Java and Bali to find work in the cities.
16. Malaysia
17. Singapore's economic output per person during this time was more than twice that of Malaysia.
18. Many businesses moved their companies to Singapore, which helped the country's economy.

Focus Skill: Make Generalizations

19. These countries are physically divided into separate parts, which complicates national unity.
20. Doing well economically supports national unity by helping people see why they want to belong to a nation.
21. The groups can have conflicts among themselves or seek independence from the country.
22. Its location has allowed it to become a major port for the region's imports and exports.
23. Industries such as telecommunications require workers with education and skills.
24. It offers economic incentives such as low tax rates.
25. The country with the higher education level will likely have the better standard of living.
26. Newspapers are unlikely to publish criticisms of the government or news that would be unfavorable to the government.

Synthesize: Answer the Essential Question

The geographic fragmentation of several Southeast Asian nations creates numerous challenges to national unity. Indonesia has attempted to relocate people from Java to the outer islands, hoping to spread the official language and help unify the country. However, this program has been far from successful, and tensions have arisen between migrants and locals. The use of Tagalog in many parts of the Philippines has helped unify the country. Singapore separated from Malaysia shortly after Malaysia gained independence from Britain, and has enjoyed tremendous economic success in the decades since then. Malaysia has focused on economic growth as a way to unify the people, and it has enjoyed stability due to decades of prosperity, but the people have been denied many political freedoms, and the government suppresses dissent in the media.

SECTION 2 STANDARDIZED TEST PRACTICE

Multiple Choice

1. C
2. B
3. D
4. A
5. B
6. D
7. C
8. A
9. B
10. D

Constructed Response

11. It served as a main transit point for sending raw material from Southeast Asia elsewhere, and for sending manufactured products to areas of Southeast Asia.
12. To: cars and machinery from the United States and Europe; From: raw materials like timber, rubber, rice, and petroleum

Extended Response

13. Singapore acts as a stopping point for raw materials travelling to other parts of the world to be used in making manufactured goods, and then serves as a port when those goods are sent back to that part of the world. This shows how the whole world cooperates to make products that the world uses.
14. Singapore's GDP more than doubled, increasing from less than $100 billion to about $220 billion.
15. Singapore's economic policies have clearly been successful, as GDP has grown as a tremendous rate over the past several decades. Other countries could invest in educating the workforce and offer incentives for foreign companies to do business there.

FORMAL ASSESSMENT

SECTION 1 QUIZ

1. C
2. B
3. D
4. A
5. D
6. A
7. Traditionally, young men live as monks for a temporary period during the rainy season. As more young people move to cities, fewer young men spend time as monks.

8. The loss of habitat threatens elephants' ability to get enough food. They wander into fields and eat crops, which brings them into conflict with the farmers who planted those crops.

SECTION 2 QUIZ

1. B	3. D	5. A
2. B	4. A	6. C

7. They are all physically fragmented countries with diverse ethnic groups.
8. The government wants an industrialized, economically prosperous society that is strictly controlled and has little crime.

SOUTHEAST ASIA TODAY TEST A

Part 1: Multiple Choice

1. A	3. A	5. C	7. B	9. D
2. D	4. B	6. B	8. C	10. A

Part 2: Interpret Maps

11. A 12. D 13. C

14. Mindanao is just as large as Luzon and has many more major cities, but it was not chosen as the location for the capital.

Part 3: Interpret Charts

15. D 16. A 17. B

18. They may be too small to consider separately. They may be close together geographically but set apart from other islands.

Part 4: Document-Based Question
Constructed Response

19. Prices might be high because the ivory trade is illegal and elephants are becoming more scarce.
20. Elephants are killed for their tusks, which is a source of ivory.
21. They have been used to perform many kinds of work.
22. The population is nearly extinct.
23. Vietnam
24. It is in danger of extinction. Although several thousand elephants remain, there used to be hundreds of thousands of elephants, so it has dropped sharply.

Extended Response Possible response:

25. Reasons include a reminder of the important role elephants have played in Southeast Asia's history and culture and the importance of keeping biodiversity to maintain balance in the environment. Recommendations include stricter laws against the ivory trade, new regulations to protect domestic elephants, and more protected areas to provide the necessary habitat for wild elephants.

SOUTHEAST EAST ASIA TODAY TEST B

Part 1: Multiple Choice

1. C	3. B	5. B	7. D	9. C
2. B	4. D	6. C	8. B	10. C

Part 2: Interpret Maps

11. D 12. A 13. B 14. B

Part 3: Interpret Charts

15. B 16. C 17. B 18. B

Part 4: Document-Based Question
Constructed Response

19. ivory
20. It threatens elephants because ivory comes from elephant tusks.
21. Thais say that elephants built their country.
22. They are not well protected. They are close to becoming extinct.
23. Myanmar
24. Vietnam and Cambodia

Extended Response Possible response:

25. The ivory trade and the lost of habitat endanger Asian elephants, as does attacks by farmers protecting their crops. As a result, elephant populations have dropped and the animal is endangered.

ACKNOWLEDGMENTS

Text Acknowledgments

210: Excerpts from *El Libertador: Writings of Simón Bolívar* by Simón Bolívar, edited by David Bushness, translated by Fred Fornoff. Copyright © 2003 by Oxford University Press. Reprinted by permission of Oxford University Press. All rights reserved.

490: Excerpts from *The Illustrated Bhagavad Gita*, translated by Ranchor Prime. Copyright © 2003 by Godsfield Press, text © by Ranchor Prime. Reprinted by permission of Godsfield Press.

542: Excerpts from *The Analects of Confucius*, translated by Simon Leys. Copyright © 1997 by Pierre Ryckmans. Used by permission of W. W. Norton & Company, Inc.

639: Data from the International Union for Conservation of Nature (IUCN) Red List of Threatened Species by IUCN. Data copyright © 2008 by the IUCN Red List of Threatened Species. Reprinted by kind permission of IUCN.

National Geographic School Publishing

National Geographic School Publishing gratefully acknowledges the contributions of the following National Geographic Explorers to our program and to our planet:

Greg Anderson, National Geographic Fellow
Katey Walter Anthony, 2009 National Geographic Emerging Explorer
Ken Banks, 2010 National Geographic Emerging Explorer
Katy Croff Bell, 2006 National Geographic Emerging Explorer
Christina Conlee, National Geographic Grantee
Alexandra Cousteau, 2008 National Geographic Emerging Explorer
Thomas Taha Rassam (TH) Culhane, 2009 National Geographic Emerging Explorer
Jenny Daltry, 2005 National Geographic Emerging Explorer
Wade Davis, National Geographic Explorer-in-Residence
Sylvia Earle, National Geographic Explorer-in-Residence
Grace Gobbo, 2010 National Geographic Emerging Explorer
Beverly Goodman, 2009 National Geographic Emerging Explorer
David Harrison, National Geographic Fellow
Kristofer Helgen, 2009 National Geographic Emerging Explorer
Fredrik Hiebert, National Geographic Fellow
Zeb Hogan, National Geographic Fellow
Shafqat Hussain, 2009 National Geographic Emerging Explorer
Beverly and Dereck Joubert, National Geographic Explorers-in-Residence
Albert Lin, 2010 National Geographic Emerging Explorer
Elizabeth Kapu'uwailani Lindsey, National Geographic Fellow
Sam Meacham, National Geographic Grantee
Kakenya Ntaiya, 2010 National Geographic Emerging Explorer
Johan Reinhard, National Geographic Explorer-in-Residence
Enric Sala, National Geographic Explorer-in-Residence
Kira Salak, 2005 National Geographic Emerging Explorer
Katsufumi Sato, 2009 National Geographic Emerging Explorer
Cid Simoes and Paola Segura, 2008 National Geographic Emerging Explorers
Beth Shapiro, 2010 National Geographic Emerging Explorer
José Urteaga, 2010 National Geographic Emerging Explorer
Spencer Wells, National Geographic Explorer-in-Residence

Photographic Credits

iii (l) ©Innovative Images/National Geographic School Publishing (c) ©Mary Lynne Ashley/National Geographic School Publishing (r) ©Martin Photography/National Geographic School Publishing. iv (l) ©Aesthetic Life Studio/National Geographic School Publishing (c) ©Gary Donnelly/National Geographic School Publishing (r) ©Cliento Photography/National Geographic School Publishing. vi Top to Bottom Left Side: ©Chris Ranier ©Gemma Atwal ©Ken Banks ©Rebecca Hale/National Geographic Stock ©Christina Conlee ©The Ocean Foundation/National Geographic Stock. Top to Bottom Right Side: ©Sybille Frütel Culhane ©Kevin Krug ©Mark Thiessen/National Geographic Stock ©Tyrone Turner/National Geographic Stock ©Adrian Jackson ©G. Anker ©Courtesy of the Jane Goodall Institute. vii Top to Bottom Left Side: ©Chris Ranier ©Rebecca Hale, National Geographic Stock ©Brant Allen ©Rolex Awards ©Beverly Joubert/National Geographic Stock. Top to Bottom Right Side: ©Beverly Joubert/National Geographic Stock ©Calit2, Erik Jepsen ©Ka'uila Barber ©National Geographic Society, Explorer Programs and Strategic Initiatives ©Sharon Farmer ©Mark Thiessen/ National Geographic Stock. viii Top to Bottom Left Side: ©Rebecca Hale/National Geographic Stock ©Lana Eklund ©Katsufumi Sato ©Victor Sanchez de Fuentes. Top to Bottom Right Side: ©Beth Shapiro ©Victor Sanchez de Fuentes ©Rachel Etherington ©David Evans/ National Geographic Society. ix ©Richard Barnes/ National Geographic Stock. x ©John Burcham, National Geographic Stock. xi ©Raul Touzon, National Geographic Stock. xii ©Rod Smith/National Geographic My Shot/National Geographic Stock. xiii ©Richard List, Corbis. xiv ©Photolibrary. xv ©David Alan Harvey/ National Geographic Stock. xvi ©Smar Jodha/National Geographic My Shot/National Geographic Stock. xvii ©Kenji Kondo/epa/Corbis. xvii ©Gavin Hellier/Alamy. xix ©Nigel Pavitt, Corbis. xx ©R. Wallace/Stock Photos/ Corbis. A1 ©Michael Dunning/Photographer's Choice/ Getty Images. A6 (bl) ©Mark Hamblin/Photolibrary (br) ©Thomas Marent/Minden Pictures/National Geographic Stock (c) ©Gerard Soury/Photolibrary (t) ©John Cancalosi/Photolibrary. A7 (bl) ©Adam Jones/Getty Images (br) ©Klaus Nigge/National Geographic Stock (c) ©DLILLC/Corbis (t) ©Thomas Marent/Minden Pictures/ National Geographic Stock. 1 (bkg) ©Franck Guiziou/ Hemis/Corbis (b, l, r) ©Mark Thiessen/National Geographic Society. 2 (bl) ©Sharon Farmer (br) ©Ka'uila Barber (cl) ©G. Anker (cr) ©Rolex Awards (tl) ©National Geographic Stock (tr) ©Calit2, Erik Jepsen. 3 (b) ©Mark Thiessen/National Geographic Stock (c) ©Ken Banks (tl)

©Gemma Atwal (tr) ©Tyrone Turner/National Geographic Stock. 4 (bkg) ©Suman Bajpeyi/National Geographic My Shot/National Geographic Stock (l) ©Jennifer Shaffer/National Geographic School Publishing (t) ©Fresco JLinga/National Geographic My Shot/National Geographic Stock. 5 ©Scott S. Warren/ National Geographic Stock. 6 (l) ©Susan Byrd/National Geographic My Shot/National Geographic Stock (r) ©Mitchell Funk/Photographer's Choice /Getty Images. 7 ©Nigel Pavitt, Corbis. 8 (bkg) ©Richard Barnes/National Geographic Stock (bl) ©Mitchell Funk/Photographer's Choice/Getty Images (cl) ©NASA Goddard Space Flight Center. (tl) ©David Evans/National Geographic Society. 10 ©Stephen Alvarez/National Geographic Stock. 11 ©Sunpix Travel/Alamy. 12 (bc) ©Mike Theiss/National Geographic Stock (cl) ©Blakeley/Alamy (cr) ©Michael S. Yamashita/National Geographic Stock 13 (tl) ©John Wark/Wark Photography, Inc. (tr) ©Mark Remaley/ Precision Aerial Photo. 15 (b) ©Michael S. Yamashita/ National Geographic Stock (t) ©Ma Wenxiao/ Sinopictures/Photolibrary. 16 ©Susan Byrd/National Geographic My Shot/National Geographic Stock. 20 ©Stephen Alvarez/National Geographic Stock. 23 ©PictureLake/Alamy. 26 ©Brooks Kraft/Corbis. 28 (bc) ©James Forte/National Geographic Stock (bl) ©Kenneth Garrett/National Geographic Stock (br) ©Michael Poliza/ National Geographic Stock (bkg) ©National Geographic Maps. 29 (bcl) ©N.C. Wyeth/National Geographic Stock (bcr) ©Justin Guarlglia/National Geographic Stock (bl) ©Kenneth Garrett/National Geographic Stock (br) ©Abraham Nowitz/National Geographic Stock 33 (b) ©David Trood/Getty Images (tr) ©Peter Carsten/National Geographic Stock. 34 ©Andrew Hasson/Alamy. 38 ©George H.H. Huey/Corbis. 41 (tc) ©Images & Volcans/ Photo Researchers, Inc. (tr) ©Chris Cheadle/Getty Images. 44 ©Bill Hatcher/National Geographic Stock. 46 (bc) ©Michael Doolittle/Alamy (cl) ©Daniel Dempster Photography/Alamy (tl) ©Tom Bean/Alamy. 47 (tl) ©Frank Krahmer/Corbis (tr) ©imagebroker/Alamy. 50 ©Panoramic Images/Getty Images. 52 ©Michael Nichols/ National Geographic Stock. 53 ©Russ Bishop/Alamy. 54 ©Kip Evans Photography. 55 ©NASA Goddard Space Flight Center. 56 (bc) ©Gordon Wiltsie/National Geographic Stock (bl) ©Ralph Lee Hopkins/National Geographic Stock (bkg) ©George Grall/National Geographic Stock (br) ©Paul Nicklen/National Geographic Stock 57 (bcl) ©Norbert Rosing/National Geographic Stock (bcr) ©William Albert Allard/National Geographic Stock (bl) ©Stuart Franklin/National Geographic Stock (br) ©Priit Vesilind/National Geographic Stock. 58 (b) ©David R. Frazier Photolibrary, Inc./Alamy (cl) ©Olivier Asselin/Alamy. 59 ©Greg Elms/ Lonely Planet Images. 63 ©John Stanmeyer/National Geographic Stock. 64 ©Nic Bothma/epa/Corbis. 65 ©Alain Nogues/Corbis. 67 ©Michael Dunning/ Photographer's Choice/Getty Images. 72 ©Romeo Gacad/ AFP/Getty Images. 74 (b) ©Tetra Images/Corbis (bkg) ©John Burcham, National Geographic Stock (c) ©Walter Meayers Edwards, National Geographic Stock (t) ©Beth Shapiro. 77 ©Petra Engle/National Geographic Stock. 78 (b) ©Phil Schermeister/National Geographic Stock (bkg) ©Menno Boermans/Aurora Photos/Corbis (t) ©SeBuKi/ Alamy. 80 ©Mike Grandmaison/Corbis. 82 ©Bill Hatcher/National Geographic Stock. 84 ©Daniel H. Bailey/Corbis. 86 (bkg) ©Jean-Pierre Lescourret/Corbis (cr) ©Mauricio Ramos. 88 ©Jon Arnold Images Ltd/ Alamy. 89 ©American School Private Collection/Peter Newark American Pictures/The Bridgeman Art Library Nationality. 90 ©Thomas Sbampato/Photolibrary. 91 ©The Granger Collection. 92 (l) ©The Granger Collection (r) ©photostock1/Alamy. 93 ©Visions LLC/Photolibrary. 95 ©William Manning/Corbis. 96 ©Bettmann/Corbis. 98 (l) ©Corbis (r) ©American School Private Collection/ Courtesy of Swann Auction Galleries/ The Bridgeman Art Library. 99 ©Corbis. 100 ©Bettmann/Corbis. 101 ©Lynn Johnson/National Geographic Stock. 102 ©The Art Archive/Museo Ciudad Mexico/Gianni Dagli Orti. 103 (l) ©Kenneth Garrett/National Geographic Stock (r) ©David R. Frazier Photolibrary, Inc./Alamy. 104 ©The Stapleton Collection/The Bridgeman Art Library. 105 ©The Stapleton Collection/The Bridgeman Art Library. 106 ©The Granger Collection. 107 (b) ©The Granger Collection (t) ©Look and Learn Magazine Ltd/The Bridgeman Art Library. 108 (l) ©North Wind Picture Archives/Alamy (r) ©Randy Faris/Corbis. 109 (l) ©Corbis (r) ©Charles & Josette Lenars/Corbis. 110 ©North Wind Picture Archives/Alamy. 111 ©Hulton Archive/Getty Images. 114 ©Joe McNally/National Geographic Stock. 116 ©Jennifer Shaffer/National Geographic School Publishing. 117 ©Mike Theiss/National Geographic Society Image Sales. 118 ©Jennifer Shaffer/National Geographic School Publishing/Art Institute of Chicago. 120 ©Car Culture/Corbis. 123 (c) ©James Forte/National Geographic Stock (tr) ©Michael Dunning/Photographer's Choice/Getty Images. 124 ©Marjorie Kamys Cotera/ Daemmrich Photography/The Image Works. 125 (b) ©Robb Kottmyer (t) ©Roger Meno. 126 ©Tono Labra/ Photolibrary. 127 ©Keith Dannemiller/Alamy. 128 ©STR/ Reuters/Corbis. 130 ©Alfredo Guerrero/epa/Corbis. 131 ©Blaine Harrington III/Alamy. 136 ©Corbis Premium RF/ Alamy. 138 (b) ©Georgios Kollidas/Alamy (bkg) ©Raul Touzon, National Geographic Stock (c) ©Stephen Alvarez, National Geographic Stock. 141 ©Konrad Wothe/Minden Pictures. 142 (b) ©Jon Arnold Images Ltd/ Alamy (bkg) ©Menno Boermans/Aurora Photos/Corbis (t) ©Danny Lehman/Corbis. 144 ©Dr. Richard Roscoe/ Visuals Unlimited, Inc. 147 ©Stuart Westmorland/ Corbis. 148 (b) ©Paul Hoekman (t) ©Bryan Wallace. 150 (bc) ©Roy Toft/National Geographic Stock (bl) ©Michael Nichols/National Geographic Stock (bkg) ©Paul Sutherland/National Geographic Stock (br) ©Steve Winter/National Geographic Society Image Sales. 151 (bcl) ©Michael Melford/National Geographic Stock (bcr) ©Bobby Haas/National Geographic Stock (bl) ©Christian Ziegler/National Geographic Stock (br) ©Roy Toft/ National Geographic Stock. 152 ©Steve Winter/National Geographic Stock. 155 ©The Bridgeman Art Library. 156 (br) ©Georgios Kollidas/Alamy (r) ©Hemis/Alamy. 157 ©Stuwdamdorp/Alamy. 158 (bl) ©Creativ Studio Heinemann/Westend61/Corbis (c) ©Interfoto/Alamy. 159 (br) ©Reuters/Corbis (tl) ©Bettmann/Corbis (tl) ©Creativ Studio Heinemann/Westend61/Corbis. 162 ©Walter Bibikow/JAI/Corbis. 164 ©Martin Gray/National Geographic Stock. 165 ©Danita Delimont/Alamy. 166 ©Nico Tondini/Photolibrary. 167 ©Rick Gerharter/Lonely Planet Images. 168 ©National Geographic Stock. 169 ©Frans Lanting/Corbis. 170 ©Photolibrary. 172 ©Yuan Man/Xinhua Press/Corbis. 173 ©Logan Abassi/UN Handout/Corbis. 174 ©JS Callahan/tropicalpix/Alamy. 175 ©Michael Dunning/Photographer's Choice/Getty Images. 176 ©Lonely Planet Images /Alamy. 178 (bkg) ©Christian Heeb/Aurora Photos (c) ©Roy Toft/National Geographic Stock (r) ©Micahel & Patricia Fogden/ Minden Pictures/National Geographic Stock. 179 ©Arterra Picture Library/Alamy. 183 (l) ©Jacques Marais/ Getty Images (r) ©Danny Lehman/Corbis. 184 ©Christian Zeigler/National Geographic Stock. 186 (b) ©Frans Lanting/Corbis (bkg) ©Rod Smith/National Geographic My Shot/National Geographic Stock (cl) ©David R. Frazier Photolibrary, Inc./Alamy (tl) ©Photograph by

Victor Sanchez de Fuentes. 189 ©Nick Gordon/Oxford Scientific (OSF)/Photolibrary. 190 (bkg) ©Menno Boermans/Aurora Photos/Corbis (bl) ©Colin Monteath/ Minden Pictures/National Geographic Stock (br) ©John Eastcott and Yva Momatiuk/National Geographic Stock. 192 ©Ivan Kashinsk/National Geographic Stock. 196 (bkg) ©Melissa Farlow/National Geographic Stock (br) ©Aldo Sessa/Tango Stock/Getty Images. 199 (br) ©Michael Nichols/National Geographic Stock (tr) ©Michael Dunning/Photographer's Choice/Getty Images. 200 ©Ethan Welty/Aurora Photos/Alamy. 201 ©Gnter Wamser/F1online digitale Bildagentur GmbH/Alamy. 202 ©Christina Conlee. 203 (b) ©Christina Conlee (t) ©Robert Clark/National Geographic Stock. 206 (bc) ©Maria Stenzel/National Geographic Stock (c) ©Peruvian School/Museo Arqueologia, Lima, Peru/Boltin Picture Library/The Bridgeman Art Library International (tr) ©McConnell, James Edwin/Private Collection /Look and Learn/The Bridgeman Art Library International. 207 ©Cro Magnon/Alamy. 208 (bc) ©The Art Archive/ Bibliothéque des Arts Décoratifs Paris/Gianni Dagli Orti (tr) ©The Art Archive/Bibliothéque des Arts Décoratifs Paris/Gianni Dagli Orti. 209 ©The Art Archive/ Kharbine-Tapabor. 211 ©Luis Marden/National Geographic Stock. 215 ©Florian Kopp/imagebroker/ Alamy. 216 ©Paolo Aguilar/epa/Corbis. 219 ©Corey Wise/Lonely Planet Images/Getty Images. 220 ©Mike Theiss/National Geographic Stock. 221 ©Richard Nowitz/National Geographic Stock. 223 ©Ivan Alvarado/ Reuters/Corbis. 224 ©Jeremy Hoare/Alamy. 225 (b) ©James P. Blair/National Geographic Stock (c) ©Nicolas Misculin/Reuters (t) ©Kit Houghton/Corbis. 226 ©Jennifer Shaffer/National Geographic School Publishing. 228 ©Keren Su/Corbis. 229 (t) ©Imaginechina/Corbis. 231 ©Robert Clark/National Geographic Stock. 232 ©Sebastiao Moreira/epa/Corbis. 234 (b) ©Mike Theiss/National Geographic Stock (bl) ©Charles Dharapak/Pool/Reuters. 240 ©Pete McBride/ National Geographic Stock. 242 (b) ©Bob Krist/National Geographic Stock (bkg) ©Richard List/Corbis (c) ©Anne Keiser/National Geographic Stock. 245 ©Atlantide Phototravel/Corbis. 246 (b) ©Yann Arthus-Bertrand/ Corbis (bkg) ©Menno Boermans/Aurora Photos/Corbis (t) ©Douglas Pearson/Corbis. 248 ©Owi-Diasign/ Photolibrary. 250 ©National Geographic Stock. 252 ©Octavio Aburto. 254 (bc) ©Anne Keiser/National Geographic Stock (bl) ©Agnieszka Pruszek/National Geographic My Shot/National Geographic Stock (bkg) ©Jim Richardson/National Geographic Stock (br) ©Steve Raymer/National Geographic Stock. 255 (bcl) ©Greg Dale/National Geographic Stock (bcr) ©James P. Blair/ National Geographic Stock (bl) ©Richard Nowitz/ National Geographic Stock (br) ©James L. Stanfield/ National Geographic Stock. 256 ©Panoramic Images/ National Geographic Stock. 258 (l) ©Richard Nowitz/ National Geographic Stock (r) ©The Art Gallery Collection/Alamy. 259 ©PoodlesRock/Corbis. 260 ©Jean-Pierre Lescourret/Corbis. 262 (l) ©Hoberman Collection/Corbis (r) ©The Bridgeman Art Library International. 263 ©North Wind Picture Archives/Alamy. 266 (bl) ©The Bridgeman Art Library (br) ©Underwood & Underwood/Corbis. 267 (bl) ©Doug Taylor/Alamy (br) ©The Bridgeman Art Library (t) ©The Bridgeman Art Library. 268 ©Richard Schlect/National Geographic Stock. 270 ©Paul Thompson/Corbis. 272 (l) ©The Bridgeman Art Library (r) ©The Gallery Collection/ Corbis. 273 (b) ©Peter Horree/Alamy (t) ©The Bridgeman Art Library. 274 ©The Bridgeman Art Library. 275 ©Scanfoto/X00729/Reuters/Corbis. 276 (l) ©Stefano Bianchetti/Corbis (r) ©Clynt Garnham/Alamy. 277 (l) ©Michael Nicholson/Corbis (r) ©Michael Nicholson/ Corbis. 279 ©DC Premiumstock/Alamy. 282 ©Rudy Sulgan/Corbis. 284 ©MARKA/Alamy. 286 (l) ©Leonardo da Vinci (1452-1519) Louvre, Paris, France/ Giraudon/ The Bridgeman Art Library (r) Claude Monet (1840-1926) Musee Marmottan, Paris, France/Giraudon/The Bridgeman Art Library Nationality. 287 ©Arnaud Chicurel/Hemis/Corbis. 288 ©The Gallery Collection/ Corbis. 289 ©Columbia/The Kobal Collection. 290 ©Jon Arnold/JAI/Corbis. 291 ©Sergiy Koshevarov/ StockPhotoPro. 292 ©Photolibrary. 294 ©Paul Seheult/ Eye Ubiquitous/Corbis. 295 ©Perutskyi Petro/ Shutterstock Photos. 296 ©Gregory Wrona/Alamy. 299 ©Michael Dunning/Photographer's Choice/Getty Images. 304 ©Grand Tour/Corbis. 306 (b) ©Gerd Ludwig/Corbis (bkg) ©Photolibrary (c) ©Gordon Wiltsie, National Geographic Stock (t) ©Rebecca Hale, National Geographic Stock. 309 ©Klaus Nigge/National Geographic Stock. 310 (bkg) ©Menno Boermans/Aurora Photos/Corbis (br) ©Bruno Morandi/Robert Harding World Imagery/Corbis (tl) ©Maxim Toporskiy/Alamy. 312 ©Denis Sinyakov/Reuters/Corbis. 314 ©Cary Wolinsky/ National Geographic Stock. 316 ©National Geographic Stock. 318 ©Gerd Ludwig/National Geographic Stock. 319 (tl) ©U.S. Geological Survey (tr) ©NASA. 320 (l) ©Sisse Brimberg/National Geographic Society (r) ©James L. Stanfield/National Geographic Society. 321 (l) ©Massimo Pizzotti/Getty (r) ©Dallas and John Heaton/ Photolibrary. 322 ©Richard Klune/Corbis. 323 ©imagebroker/Alamy. 324 ©The Bridgeman Art Library (r) ©Cary Wolinsky/National Geographic Society. 325 (l) ©The Bridgeman Art Library. 326 ©North Wind Picture Archives/Alamy. 328 (l) ©Bettmann/Corbis (r) ©The Art Archive. 329 (b) ©Bettmann/Corbis (t) ©Thomas Johnson/Sygma/Corbis. 332 ©Paul Harris/JAI/Corbis. 334 ©Arne Hodalic/Corbis. 335 (tc) ©Michael Runkel/Robert Harding World Imagery/Corbis (tr) ©Maria Stenzel/ National Geographic Stock (tr) ©Sean Sprague/ Photolibrary. 336 ©Olaf Meinhardt/Visum/Fotofinder. 339 ©Kristel Richard/Grand Tour/Corbis. 340 ©Shepard Sherbell/Corbis Saba. 342 ©Imagesource/Photolibrary. 344 ©Oleg Nikishin/Stringer/Getty Images. 347 (c) ©iStockphoto ©Michael Dunning/Photographer's Choice/Getty Images. 352 ©iStockphoto. 354 (b) ©Ingo Arndt/Minden Pictures/National Geographic Stock (bkg) ©David Alan Harvey/National Geographic Stock (c) ©Mitsuaki Iwago/Minden Pictures/National Geographic Stock (t) ©Kakenya Ntaiya. 357 ©Top-Pics TBK/Alamy. 358 (bkg) ©Menno Boermans/Aurora Photos/Corbis (bl) ©tbkmedia/Alamy (tr) ©Michael Poliza/National Geographic Stock. 360 ©Michael Nichols/National Geographic Stock. 362 ©Philippe Bourseiller/Getty Images. 364 ©Ian Nichols/National Geographic Stock. 366 ©Mike Hutchings/Reuters. 368 (bc) ©Beverly Joubert/National Geographic Stock (bl) ©Beverly Joubert/ National Geographic Stock. 370 ©Gerald Hoberman/ Hoberman Collection UK/Photolibrary. 372 (bl) ©The Trustees of the British Museum/Art Resource (br) ©ADB Travel/dbimages/Alamy. 373 ©HIP/Art Resource. 376 ©Private Collection/Look and Learn/The Bridgeman Art Library International. 377 (bl) ©Mary Evans Picture Library/The Image Works (br) ©Bruce Dale/National Geographic Stock. 378 (bc) ©James L. Stanfield/National Geographic Stock (bl) ©Tim Laman/National Geographic Stock (bkg) ©George Steinmetz/National Geographic Stock (br) ©Annie Griffiths/National Geographic Stock. 379 (bcl) ©Roy Toft/National Geographic Stock (bcr) ©Tino Soriano/National Geographic Stock (bl) ©Jodi Cobb/National Geographic Stock (br) ©Ed Kashi/ National Geographic Stock. 382 ©Ralph Lee Hopkins/

National Geographic Stock. 384 ©Vanessa Burger/Images of Africa Photobank Alamy. 386 ©Paul Gilham–FIFA/ FIFA via Getty Images. 387 ©David Alan Harvey/ National Geographic Stock. 388 (bc) ©Nigel Pavitt/John Warburton-Lee Photography/Alamy (bl) ©Sean Sprague/ Still Pictures/Photolibrary. 389 (cl) ©Michael Nichols/ National Geographic Stock (tr) ©Suzi Eszterhas/Minden Pictures/National Geographic Stock. 390 ©Jane Goodall Institute. 391 (bkg) ©Gerry Ellis/ Minden Pictures/ National Geographic Stock (br) ©Wade Davis/Ryan Hill. 392 ©Finbarr O'Reilly/Reuters. 394 ©George Steinmetz/ Corbis. 396 (b) ©Pascal Maitre/National Geographic Stock (tr) ©Joerg Boethling/Alamy. 398 ©Louise Gubb/ Corbis. 399 ©Michael Dunning/Photographer's Choice/ Getty Images. 400 ©Frederic Courbet/Still Pictures/ Photolibrary. 402 ©Ulrich Doering/Alamy. 403 ©Trinity Mirror/Mirrorpix/Alamy. 406 (bl) ©Chris Stenger/FN/ Minden Pictures/National Geographic Stock (br) ©Walker, Lewis W./National Geographic Stock. 407 (bkg) ©Clement Philippe/Arterra Picture Library/Alamy (bkg) ©Tim Fitzharris/Minden Pictures/National Geographic Stock (cf) ©Mattias Klum /National Geographic Stock (l) ©Tom Vezo/Minden Pictures/National Geographic Stock (rbkg) ©Ted Wood/Aurora Photos (rf) ©Thomas Lehne/ Alamy. 408 ©Anup Shah/Corbis. 410 (b) ©Martin Gray/ National Geographic Stock (bkg) ©Smar Jodha/National Geographic My Shot/National Geographic Stock (c) ©David Boyer/National Geographic Stock (t) ©Thomas Culhane. 413 ©Vanessa Lefort/National Geographic My Shot/National Geographic Stock. 414 (bkg) ©Menno Boermans/Aurora Photos/Corbis (cl) ©Gary Cook/Alamy (cr) ©Peter Adams/Getty Images. 416 ©Ed Kashi/National Geographic Stock. 418 ©Fischer Gunter/WoodyStock/ Alamy. 420 ©Chris Bradley/Axiom/photolibrary. 422 (bl) ©Victor R. Boswell, Jr/National Geographic Stock (br) ©Scala/Art Resource 423 (bl) ©Erich Lessing/Art Resource (br) ©Corbis. 424 ©Richard Nowitz/National Geographic Stock. 425 ©Oliver Weiken/epa/Corbis. 426 ©Bachmann Bachmann/F1 Online/photolibrary. 427 (bl) ©James Brunker/Alamy (tr) ©Kordcom Kordcom/age fotostock/photolibrary. 428 ©Yann Arthus-Bertrand/ Corbis. 429 ©The Art Archive/Topkapi Museum Istanbul/Dagli Orti. 430 ©Gwill Owen/Sylvia Cordaiy Photo Library Ltd /Alamy. 431 ©NASA/JSC/Gateway to Astronaut Photography of Earth. 432 (bl) ©Paul Sutcliffe/ Alamy. (br) ©Erich Lessing/Art Resource. 433 (bc) ©Mary Jelliffe/Ancient Art & Architecture Collection Ltd./ Alamy (tr) ©Kenneth Garrett/National Geographic Stock. 434 ©Kenneth Garrett/National Geographic Stock. 438 ©Keren Su/Corbis. 440 ©Alberto Arzoz/Axiom/Aurora Photos. 441 ©Walter Bibikow/Jon Arnold Images Ltd./ Alamy. 442 ©Gavin Hellier/Alamy. 443 ©David Bathgate/ Corbis. 444 (b) ©Radius Images/Corbis (bl) ©G. Anker. 445 ©Hanan Isachar/Corbis. 446 ©NASA/Science Faction/Corbis. 447 ©Matthias Seifert/Reuters/Corbis. 449 (tr) ©Umit Bektas/Reuters/Corbis ©Felipe Trueba/ epa european pressphoto agency. 450 ©Mohammad Berno/Document Iran/Corbis. 452 ©David Rubinger/ Time & Life Pictures/Getty Images. 454 ©Sabah Arar/ AFP/Getty Images. 457 ©Tim Gurney/Alamy. 459 ©Shehzad Noorani/Stillpictures/Aurora Photos. 463 ©Mark Thiessen/National Geographic Stock. 464 ©Bildarchiv Preussischer Kulturbesitz/Art Resource. 466 (b) ©Peter Adams/Corbis (bkg) ©Tibor Bognar/ agefotostock (c) ©Kenji Kondo/epa/Corbis (t) ©Hussain RAE photos. 469 ©Lynn M. Stone/Nature Picture Library. 470 (bkg) ©Menno Boermans/Aurora Photos/Corbis (bl) ©Tiziana and Gianni Baldizzone/Corbis (l) ©Dinodia Images/Alamy (tc) ©Stephen Sharnoff/National Geographic Stock. 472 (bkg) ©Bobby Model/National Geographic Stock (l) ©James L. Stanfield/National Geographic Stock. 474 ©Frederic Soltan/Sygma/Corbis. 476 ©Lynsey Addario/National Geographic Image Collection. 478 ©Prakash Singh/AFP/Getty Image. 479 ©Blue Legacy International. 481 ©Michael Dunning/ Photographer's Choice/Getty Images. 482 ©Luca Tettoni/ Corbis. 483 (b) ©The Schoyen Collection (t) ©The Schoyen Collection. 484 (l) ©Silvio Fiore/SuperStock (r) ©The Trustees of the British Museum. 485 ©Thomas Retterath/Getty Images. 488 ©Jeremy Horner/Corbis. 489 ©Lineair/Photolibrary. 490 ©Bettmann/Corbis. 491 ©Art Directors & TRIP/Alamy. 494 ©Harish Tyagi/epa/Corbis. 496 ©Louise Batalla Duran/Alamy. 497 ©Bruce Dale/ National Geographic Stock. 498 (bkg) ©Ed Kashi/ National Geographic Stock (l) ©Jodi Cobb/National Geographic Stock. 500 ©Ajay Verma/Reuters/Corbis. 501 (b) ©Foodfolio–StockFood Munich (t) ©Abraham Nowitz/National Geographic Stock. 502 (bkg) ©David Cumming/Eye Ubiquitous/Corbis (bl) ©Robert Wallis/ Corbis (bc) ©Stephen Romilly/Alamy. 503 (b) ©Dharma Productions/The Kobal Collection (t) ©Frazer Harrison/ Getty Images. 504 ©Eric Feferberg/Pool/Reuters. 506 ©Fredrik Renander/Alamy. 507 (t1) ©Ed Kashi/National Geographic Stock (t2) ©Ed Kashi/National Geographic Stock (t2) ©Fridmar Damm/Corbis (t3) ©Akhtar Soomro/ Deanpictures/The Image Works (t4) ©Akhtar Soomro/ Deanpictures/The Image Works (t5) ©Andrew Holbrooke/ Corbis (t7) ©National Geographic Maps (t8) ©National Geographic Maps. 508 ©Ed Kashi/National Geographic Image Collection. 510 ©Sajjad Hussain/AFP/Getty Images. 511 ©Dinodia Photo Library. 512 ©Akhtar Soomro/Deanpictures/The Image Works. 514 ©Andrew Holbrooke/Corbis. 522 (b) ©Michael Nichols/National Geographic Stock (bkg) ©George Steinmetz/National Geographic Stock (c) ©Gavin Hellier/Alamy (t) ©Albert Lin. 525 ©Mitsuaki Iwago/Minden Pictures/National Geographic Stock. 526 (b) ©Chun Ki Leung/National Geographic My Shot /National Geographic Stock (bkg) ©Menno Boermans/Aurora Photos/Corbis (t) ©Wang Jianjun/TAO Images Limited/Alamy. 528 ©Fritz Hoffmann/National Geographic Stock. 530 ©Reuters/ Mainichi Shimbun. 532 ©Toby Adamson/Axiom Photographic Agency/Getty Images. 534 ©Alison Wright/ National Geographic Stock. 536 (b) ©Unterthiner, Stefano/National Geographic Stock (cl) ©Katsufumi Sato/National Geographic Society. 538 (bl) ©Richard Swiecki/Royal Ontario Museum/Corbis (br) ©Ira Block/ National Geographic Stock. 539 (bc) ©O. Louis Mazzatenta/National Geographic Stock (tc) ©Atlantide Phototravel/Corbis. 540 (bc) ©Michael S. Yamashita/ National Geographic Stock (bl) ©Justin Guariglia/ National Geographic Stock (bkg) ©O. Louis Mazzatenta/ National Geographic Stock (br) ©Michael S. Yamashita/ National Geographic Stock. 541 (bcl) ©Kenneth Ginn/ National Geographic Stock (bcr) ©Kate Staszczak/ National Geographic My Shot/National Geographic Stock (bl) ©Michael S. Yamashita/National Geographic Stock (br) ©Ira Block/National Geographic Stock. 542 ©Shiwei/ Best View Stock/photolibrary. 544 ©Redlink/Corbis. 546 (bkg) ©Gregory A. Harlin/National Geographic Stock (b) ©National Geographic Maps. 548 ©Wendy Connett/ Alamy. 549 ©Rob Howard/Corbis. 550 ©Ira Block/ National Geographic Stock. 551 ©Ira Block/National Geographic Stock. 552 ©Asian Art & Archaeology, Inc./ Corbis. 553 (b) ©Private Collection/Peter Newark Military Pictures/The Bridgeman Art Library International. 554 (bl) ©H. Edward Kim/National Geographic Stock (br) ©Korea News Service/Reuters/ Corbis. 555 (bl) ©John Van Hasselt/Sygma/Corbis ©(br) ©The Trustees of The British Museum/Art Resource. 556

(bl) ©Bettmann/Corbis. 557 (l) ©Bruce Burkhardt/Corbis (r) ©Michael S. Yamashita/National Geographic Stock. 560 ©B.S.P.I./Corbis. 562 ©Buena Vista Images/Getty Images. 564 (bl) ©The Trustees of the British Museum/Art Resource. 565 ©View Stock/Alamy. 566 ©Justin Guariglia/National Geographic Stock. 568 (bc) ©Mark Leong/National Geographic Stock (bl) ©Richard Nowitz/National Geographic Stock (bkg) ©Scott S. Warren/National Geographic Stock (br) ©Jason Teale/National Geographic My Shot/National Geographic Stock. 569 (bcl) ©Pete Ryan/National Geographic Stock (bl) ©Michael S. Yamashita/National Geographic Stock (bcr) ©Thomas J. Abercrombie/National Geographic Stock (br) ©Jodi Cobb/National Geographic Stock. 570 (bl) ©Everett Kennedy Brown/epa/Corbis (cl) ©Everett Kennedy Brown/epa/Corbis. 571 ©Studio Ghibli/Tokuma Shoten/The Kobal Collection. 572 (b) ©Jose Fuste Raga/Corbis. 574 (b) ©Jianan Yu/Reuters/Corbis. 577 ©Michael Dunning/Photographer's Choice/Getty Images. 578 ©epa/Corbis. 580 (bl) ©Nicky LohReuters/Corbis (bkg) ©Justin Guariglia/National Geographic Stock. 582 ©Asia File/Alamy. 585 (bkg) ©SJ. Kim/Flickr/Getty Images (tr) ©Jane Sweeney/JAI/Corbis. 586 (bc) ©Phil Iossifidis/National Geographic My Shot/National Geographic Stock (bl) ©Paul Chesley/National Geographic Stock (bkg) ©Alicia Pudsey/National Geographic My Shot/National Geographic Stock (br) ©Arkadiusz Dudzinski/National Geographic My Shot/National Geographic Stock. 592 ©John Woodcock/iStock Vectors/Getty Images. 586 (bcl) ©Winfield Parks/National Geographic Stock (bcr) ©Paul Chesley/National Geographic Stock (bl) ©Martin Gray/National Geographic Stock (br) ©Justin Guariglia/National Geographic Stock 593 ©Natureworld/Alamy. 594 (b) ©Paul Chesley/National Geographic Stock (bkg) ©Nigel Pavitt/Corbis (c) ©John Stanmeyer LLC, National Geographic Stock (t) ©Kevin Krug. 597 ©Tui De Roy/Minden Pictures/National Geographic Society Image Sales. 598 (bkg) ©Menno Boermans/Aurora Photos/Corbis (bl) ©NASA (tr) ©Mary Plage/Photolibrary. 600 ©Brant Allen. 602 (bkg) ©Neil Rabinowitz/Corbis (bl) ©James P. Blair/National Geographic Society Image Sales. 604 ©John Stanmeyer/National Geographic Society Image Sales. 606 ©Tim Laman/National Geographic Society Image Sales. 607 (bkg) ©Tim Laman/National Geographic Society Image Sales (cl) ©Tim Laman/National Geographic Society Image Sales (cr) ©Tim Laman/National Geographic Society Image Sales. 608 (c) ©Robert Clark/National Geographic Society Image Sales (cb) ©Robert Harding Picture Library Ltd/Alamy. 609 (br) ©View Stock/Alamy. 612 (bl) ©Stringer/Indonesia/Reuters/Corbis (r) ©Lindsay Hebberd/Corbis. 613 ©FPG/Hulton Archive/Getty Images. 615 ©W.E. Garrett/National Geographic Stock. 618 ©Jose Fuste Raga/Corbis. 620 (b) ©James Nachtwey VII/National Geographic Society Image Sales. 621 (tl) ©Friedrich Stark/Alamy (tr) ©Robert Clark/National Geographic Society Image Sales. 622 (bkg) ©Danita Delimont/Gallo Images/Getty Images (bl) ©Jon Arnold Images Ltd/Alamy. 623 (cr) ©WoodyStock/Alamy. 624 ©Photolibrary. 625 ©National Geographic School Publishing. 626 (b) ©Ulet Ifansasti/Getty Images. 627 (tl) ©Buddy Mays/Alamy. 628 ©Imagemore Co., Ltd./Corbis. ©AFP/Getty Images 629 ©AFP/Getty Images. 630 ©Peter Adams/JAI/Corbis. 631 ©John Stanmeyer/National Geographic Society Image Sales. 633 (c)© Lauca Images/Alamy (t) ©Michael Dunning/Photographer's Choice/Getty Images. 634 ©Tengku Mohd Yusof/Alamy. 639 (b) ©Theo Alofs/Minden Pictures/National Geographic Stock (cl) ©Suzi Eszterhas/Minden Pictures/National Geographic Stock (cr) ©John Cancalosi/Photolibrary (tl) ©Tom Brakefield/Photolibrary (tr) ©Top-Pics TBK/Alamy. 642 (b) ©R. Wallace/Stock Photos /Corbis (bkg) ©Yva Momatiuk & John Eastcott/Minden Pictures/National Geographic Stock (c) ©Pixtal Images/Photolibrary (t) ©Ka'uila Barber. 645 ©Mitsuaki Iwago/Minden Pictures/National Geographic Stock. 646 (bkg) ©Menno Boermans/Aurora Photos/Corbis (bl) ©Fred Bavendam/Minden Pictures/National Geographic Stock (tr) ©Michel Renaudeau/Photolibrary. 648 ©Kevin Schafer/Alamy. 649 ©Look Die Bildagentur der Fotografen GmbH/Alamy. 650 (b) ©Martin Harvey/Corbis (t) ©John Carnemolla/Corbis. 652 ©Brian Skerry/National Geographic Society Image Sales. 655 (t) ©Michael Dunning/Photographer's Choice/Getty Images (c) ©Jeff Hunter/Getty Images. 656 ©Catherine Karnow/Corbis. 657 ©Belinda Wright/National Geographic Society Image Sales. 658 (b) ©Rob Howard/Corbis (t) ©Walter Meayers Edwards/National Geographic Society Image Sales. 659 (c) ©Jean Leo Dugast/Sygma/Corbis (t) ©Image Source/Corbis. 660 ©Patrick Eden/Alamy. 661 (bl) ©Corbis (l) ©Caro/Alamy (t) ©The Bridgeman Art Library. 662 ©Chris Ranier. 663 ©Chris Ranier. 666 ©Massimo Ripani/Grand Tour/Corbis. 668 ©Amy Toensing/National Geographic Stock. 669 ©David Ball/Alamy. 670 ©Stringer/Australia/X01245/Reuters/Corbis. 672 ©Randy Olson/National Geographic Stock. 674 ©George F. Mobley, National Geographic Stock. 676 ©Bates Littlehales/National Geographic Stock. 678 ©POOL/Reuters/Corbis. 679 ©Phil Walter/Getty Images. 681 (bkg) ©David Doubilet/National Geographic Stock (tl) ©David McLain/National Geographic Stock (tr) ©David Wall/Alamy. 682 ©David L. Moore/Alamy. 683 ©Tui De Roy/Minden Pictures/National Geographic Stock. 688 ©Paul Nicklen/National Geographic Stock. R39 ©Fresco JLinga/National Geographic My Shot/National Geographic Stock. R40 ©George Grall/National Geographic Stock. R41 ©SERDAR/Alamy. R42 ©Bettmann/Corbis. R44 ©Hulton Archive/Getty Images. R45 ©Svabo/Alamy. R47 ©Scott Olson/Getty Images. R49 ©Lordprice Collection/Alamy. R50 ©Parbul TV via Reuters TV/Reuters/Corbis. R51 ©Markus Altmann/Corbis. R52 (b) ©Mario Lopez/epa/Corbis (t) ©Alinari Archives/Corbis. R53 (b) ©Anindito Mukherjee/epa/Corbis (t) ©Werner Forman/Art Resource. R54 (b) ©Ronen Zvulun/Reuters/Corbis (t) ©The Art Archive/Museo del Prado Madrid. R55 (b) ©Mak Remissa/epa/Corbis (t) ©Rubin Museum of Art/Art Resource. R56 (b) ©Joe McNally/National Geographic Stock (t) ©Biju/Alamy. R57 (b) ©Robert Harding World Imagery/Corbis (t) ©Art Directors & TRIP/Alamy. R58 (b) ©Christian Kober/Photolibrary (t) ©National Palace Museum Taiwan/The Art Archive.

Map Credits

Mapping Specialists, LTD., Madison, WI.
National Geographic Maps, National Geographic Society

Illustrator Credits

Precision Graphics

R128 ACKNOWLEDGMENTS

ACKNOWLEDGMENTS | TEACHER'S EDITION

Text Acknowledgments

RB57: Excerpt from "Bethlehem 2007 A. D." by Michael Finkel from *National Geographic* December 2007. Copyright © Michael Finkel. Reprinted by permission of the author.

Photographic Credits

603 ©Anuar Ahmad/123rf RB10 © Suthep Kritsanavarin/epa/Corbis. 627 ©Gerry Ellis/Minden Pictures/National Geographic Stock. RB16 © Classic Image/Alamy RB19 (cl) ©Tim Laman/National Geographic Stock (l) ©Tim Laman/National Geographic Stock (llc) ©Tim Laman/National Geographic Stock (rrc) ©Tim Laman/National Geographic Stock cr ©Tim Laman/National Geographic Stock r ©Tim Laman/National Geographic Stock. RB32 ©Jack Birns/Time Life Pictures/Getty Images RB39 ©Robert Clark/National Geographic Society Image Sales RB82 ©Rebecca Hale/National Geographic Stock RB87 ©Rebecca Hale/National Geographic Stock.

Map Credits

Mapping Specialists, LTD., Madison, WI.
National Geographic Maps, National Geographic Society

Illustrator Credit

Precision Graphics

Front Cover

(bkg) © Scott Kemper / Alamy (front)©Lonely Planet Images/Getty Images